Cybersecurity Operations and Fusion Centers

Cybersecurity Operations and Fusion Centers: A Comprehensive Guide to SOC and TIC Strategy by Dr. Kevin Lynn McLaughlin is a must-have resource for anyone involved in the establishment and operation of a Cybersecurity Operations and Fusion Center (SOFC). Think of a combination cybersecurity SOC and cybersecurity Threat Intelligence Center (TIC). In this book, Dr. McLaughlin, who is a well-respected cybersecurity expert, provides a comprehensive guide to the critical importance of having an SOFC and the various options available to organizations to either build one from scratch or purchase a ready-made solution. The author takes the reader through the crucial steps of designing an SOFC model, offering expert advice on selecting the right partner, allocating resources, and building a strong and effective team. The book also provides an in-depth exploration of the design and implementation of the SOFC infrastructure and toolset, including the use of virtual tools, the physical security of the SOFC, and the impact of COVID-19 on remote workforce operations. A bit of gamification is described in the book as a way to motivate and maintain teams of high-performing and well-trained cybersecurity professionals.

The day-to-day operations of an SOFC are also thoroughly examined, including the monitoring and detection process, security operations (SecOps), and incident response and remediation. The book highlights the significance of effective reporting in driving improvements in an organization's security posture.

With its comprehensive analysis of all aspects of the SOFC, from team building to incident response, this book is an invaluable resource for anyone looking to establish and operate a successful SOFC. Whether you are a security analyst, senior analyst, or executive, this book will provide you with the necessary insights and strategies to ensure maximum performance and long-term success for your SOFC. By having this book as your guide, you can rest assured that you have the knowledge and skills necessary to protect an organization's data, assets, and operations.

Security, Audit and Leadership Series

Series Editor: Dan Swanson, Dan Swanson and Associates, Ltd., Winnipeg, Manitoba, Canada.

The **Security, *Audit* and Leadership Series** publishes leading-edge books on critical subjects facing security and audit executives as well as business leaders. Key topics addressed include Leadership, Cybersecurity, Security Leadership, Privacy, Strategic Risk Management, Auditing IT, Audit Management and Leadership

Cybersecurity Operations and Fusion Centers
A Comprehensive Guide to SOC and TIC Strategy

Kevin Lynn McLaughlin

CRC Press
Taylor & Francis Group
Boca Raton London New York

CRC Press is an imprint of the
Taylor & Francis Group, an **informa** business

Cover Image Credit: Shutterstock

First edition published 2024
by CRC Press
6000 Broken Sound Parkway NW, Suite 300, Boca Raton, FL 33487-2742

and by CRC Press
4 Park Square, Milton Park, Abingdon, Oxon, OX14 4RN

ISBN: 9781032194356 (hbk)
ISBN: 9781032194363 (pbk)
ISBN: 9781003259152 (ebk)

DOI: 10.1201/9781003259152

Typeset in Caslon
by Newgen Publishing UK

Thank you to my friend and colleague Ibn Akbar at Nice Touch Editing Services for helping me with the edits for this book.

Thank you to my friend and peer Gulshan Verma in Hyderabad, India for his help, assistance, and insights on Chapters 1 and 2.

Lastly, thank you to all the cybersecurity professionals I've been lucky enough to work with over my long career.

Contents

PART III REPORTING AND METRICS

PART IV LEADERSHIP ALIGNMENT AND SUPPORT

Preface

My friends—and if you are a member of this profession that I am most passionate about then you are indeed my friend—it is a great pleasure to welcome you to this insightful guide on the complexities and nuances of Cybersecurity Operations and Fusion Center (SOFC) deployment. This book has been crafted with the utmost care and precision to provide you with an in-depth understanding of the critical components involved in building and deploying a successful SOFC. In the pages that follow, you will embark on a journey that begins with a comprehensive examination of the why and the how of building an SOFC. This journey will take you through the essential design and building processes involved in creating an SOFC from the ground up. You will learn about the importance of building a core team and the key components involved in designing the model. The journey will then move on to explore the various tools and operations that are critical to the successful functioning of an SOFC. You will gain an in-depth understanding of the infrastructure and toolset that are required to support SOFC operations, as well as an exploration of the vital role played by SecOps and Detection, Response, and Remediation in ensuring the security of the organization. In the final stages of the journey, you will delve into the world of reporting and metrics, where you will learn about the crucial role that reporting and metrics play in enabling organizations to evaluate and optimize the performance of their SOFC. You will also explore the critical importance of leadership alignment and support in ensuring the success of an SOFC, as well as the key components of a turnkey solution. Finally, you will find useful appendices that contain templates, providing you with some resources to support your continued learning and professional development in the field of SOFC deployment.

That being said, it is time to tighten your cogs and prepare for an adventure of a lifetime! As we embark on this journey of intrigue and discovery, let us cast off the moorings and set sail towards uncharted territories of the mind! The journey ahead shall be filled with the marvels of cybersecurity, where algorithms, cryptographic protocols, and the hum of servers shall transport us to realms beyond our greatest expectations! Let us navigate the digital landscapes and secure our systems from the dangers that lurk in the shadows. As we traverse this digital frontier, we must be vigilant and ensure that the cogs of our security measures are well oiled and functioning at peak efficiency. So fasten your seat belts, my dear colleagues, and let us embark on this journey together!

About the Author

Kevin Lynn McLaughlin, PhD, CISO, CISM, CISSP, PMP, ITIL Master, LSSBB, GIAC-GSLC, CRISC, is a highly accomplished cybersecurity expert with a diverse background in law enforcement, corporate security, and cybersecurity. He proudly served in the U.S. Army and was a U.S. Special Agent before making a significant impact in the world of corporate security. With over 39 years of experience in the field, Dr. McLaughlin has demonstrated his expertise in creating and leading three Global Cybersecurity Programs for Fortune 300 companies, establishing Global Security Operations Centers, and designing and implementing a Global Cybersecurity Architecture. He is a veteran in global cyber investigations, having led over 800 investigations, and is a skilled executive manager who has led Global Cyber and Corporate Security teams. Kevin is a highly sought after speaker, having spoken at RSA, and has advised Board of Directors on various cybersecurity topics. He is also an expert in executive protection and securing critical manufacturing, manufacturing, consumer goods, and healthcare environments.

PART I
Building and Deployment

1

What Is a Cybersecurity Operations and Fusion Center (SOFC) and Why Do You Need One?

A successful cybersecurity professional must strive to simplify the complex realm of cybersecurity and make it easier to navigate and understand.

Dr. Kevin Lynn McLaughlin, PhD

Imagine a central defense system that focuses on monitoring, defending, and communicating awareness of existing/mitigated cyber threats. A place where digital wars are fought daily. A place where one might hear a group of security experts applaud their victory over adversaries during times of crisis. A one-of-a-kind location where routine day-to-day cybersecurity operations are conducted. This place is known as a Cybersecurity Operations and Fusion Center (SOFC). But before we go any further about that concept, let's first take a step back and talk about a Cybersecurity Security Operations Center (SOC).

In its simplest form, an SOC is a facility full of security experts that are dedicated to defending an organization's data and assets by continuously monitoring the company's security posture. The SOC team's key objective is to track, evaluate, and respond to cybersecurity events by using specialized tools and technologies, and a strong collection of security processes that align with both cybersecurity and business requirements. An SOC is usually filled with cybersecurity officers, engineers, incident responders, investigators, and administrators. Most SOC operations consist of strong, cross-functional efforts between various security response teams. These teams work collaboratively to respond to and mitigate observed security issues that arise on a day-to-day basis within the organization.

DOI: 10.1201/9781003259152-2

Before we dive deep into how an SOC and an SOFC are different, let us consider a fictional organization named NEXTCORP. NEXTCORP is a multimillion-dollar e-commerce business with a global presence in 300 locations worldwide. This company sells a wide range of consumer products through its e-commerce portal. NEXTCORP plans to integrate a social network application and other associated applications with its e-commerce website. This integration will enable consumers to not only connect with each other, but also help them engage with NEXTCORP's customer service and other support services. Integrating these applications will also allow customers to subscribe to company news, new products, subscriptions, magazines, etc. The service will be deployed on NEXTCORP's cloud storage networks and will utilize different third-party services for a variety of needs. These needs include translating photographs of contracts into PDF files or improving the images and quality of content posted on the social network. NEXTCORP has a team of four security professionals, including an administrator. However, the organization has no dedicated SOC. The current cybersecurity staff uses a few cybersecurity-related processes to protect its technology business and defend it from basic security issues and internal escalations. The staff monitors password changes, new employee account creation, account lockouts, etc. As NEXTCORP grows into a global company it realizes that it needs to expand its cybersecurity capabilities. The company recognizes that it needs to be properly staffed to defend against cybersecurity attacks 24×7. NEXTCORP decides to grow its cybersecurity capabilities through the creation of an SOC.

SOCs have been around for a while now. The earliest were in operation around the late 1990s to early 2000s. In comparison to other corporate functions, this is a new phenomenon for organizations to build and put into operation. Since the inception of SOCs, its primary goal has been to protect their organization's data, assets, and ability to conduct business in a way that ensures stakeholder and where relevant, shareholder confidence. SOCs vary in size. The size is often determined by how large or financially set the organization is that the SOC is defending. An SOC is typically composed of various tools that allow the team to monitor, detect, and respond to security alerts, and tools that enable the team to contain the event once it is detected and responded to. An SOC tracks and evaluates events on networks, routers, endpoints, directories, software, websites, and other technology-enabled systems. SOCs search for irregularities that may be evidence of a security incident or intrusion. The SOC is responsible for ensuring that cybersecurity threats and associated incidents are timely detected, assessed, analyzed, reviewed, defended, and archived.

There are plenty of tools available that can be used by the SOC analyst to accomplish these tasks. Some tools consist of machine learning methodologies to analyze substantial amounts of data quickly, determine abnormal data patterns quickly, and send off relevant alerts to the monitoring team so that proper action can take place. These tools start off by collecting activity logs from across the infrastructure and storing them into a specific location or a single Security Analytic System (SAS). There are a lot of great tools to choose from. Here are a few common SASs as referenced by Gartner: Securonix, RSA, FireEye, McAfee, AT&T, and Fortinet. An internet search will find many more SAS that can be successfully used by the SOC to monitor and detect potential security events.

NEXTCORP is now aware of what an SOC is and how it operates, but they recently heard the term Fusion Center. They are wondering what a Fusion Center is and whether they should leverage Fusion Center methods into their planned future SOC operations. Fusion Centers came into prominent focus in the United States of America (USA) after the 9/11 attacks. The USA Department of Homeland Security (DHS) established Fusion Centers across the United States. A Fusion Center is designed to gather, analyze, and make active use of intelligence that is relevant to defending the homeland or an organization. By taking in and analyzing intelligence data and comparing the data for relevancy to systems and assets in their organization, the SOFC can proactively identify and classify risks and threats, as well as drive remediation of potential issues. DHS has funded and recommended the creation of Fusion Centers across federal, state, and local governments. These Fusion Centers are designed to take disparate pieces of information on a variety of subjects and "fuse" them together to be able to recognize indicators of potential threats and risks. These centers are staffed with personnel who have intelligence and law enforcement backgrounds. The staff analyzes data to determine the threat or risk associated with the received intelligence. Due to the current threat landscape, the number of cyber defenders deployed by DHS and other government agencies is growing by large numbers every week. Many of these new cybersecurity professionals are working on Fusion Center operations to help take proactive action against cyber threats.

The concept of gathering intelligence data and proactively assessing the threat and risk that is associated with the data is of paramount importance to providing effective cybersecurity for an organization. It is important that organizations stay in front of potential attacks because cyberattacks from organized crime and government actors continue to spread. According to the Verizon 2020 Data Breach Report, organized crime accounted for 55% of known data breaches. By combining this Fusion Center concept and related work-streams with the traditional SOC, we produce an effective layer of cybersecurity protection that is greater than what most organizations currently deploy.

An SOC when combined with a Fusion Center becomes a hybrid model that I will refer to as the SOFC. This hybrid approach combines the working operations of the SOC with the working operations of a Fusion Center and becomes a best-in-class approach for organizations that want to stay ahead of cyber threats. The SOFC will be responsible for traditional SOC activities such as monitoring, detecting, and responding to security alerts as well as the tools that enable the team to contain an event once it is detected and responded to. Additionally, the SOFC gathers external and internal intelligence data that can be proactively correlated against the corporation's internal systems for relevancy, risk, and threat rating. This intelligence component needs to adhere to best practices promoted across the intelligence communities. When these functions are combined, the cybersecurity team can review alerts coming in, triage those alerts, and compare them to known indicators of compromises and known patterns associated with cyberattacks. The team is also able to proactively identify and categorize potential threats before they cause a negative impact to their corporate systems. The identification and categorization activity is followed by analyzing the information for relevancy within the environment. Example: The SOFC intelligence analyst sees an interesting alert about Acme Database version 3.1 across one of the intelligence feeds that an organization subscribes to. The alert mentions that this database has a known vulnerability that is being actively exploited. The analyst takes that information and checks across their corporate infrastructure to see if anyone is using Acme Database version 3.1. In our example, we will assume that the analyst finds two core database systems running the vulnerable database. The analyst now factors into their intelligence report (see the appendix for a sample report) that the Acme Database version 3.1 is vulnerable and there is an active exploit for it. Their company has this database in use and determines that there is a high threat/risk to the corporation. The threat/risk report is completed, along with recommendations for remediation. The SOFC intelligence manager forwards the report via email to the owners of the vulnerable databases with a note to expedite their remediation as they are in near-term danger of being compromised. It is good practice to follow up the email with a phone call to the owner letting them know what the risk is and why it is important for them to take immediate remediation action, and to offer the help of the SOFC support members if needed.

Based on the interpretation of present-day data, the Fusion Center cybersecurity analysts need to have a natural curiosity that allows them to have the foresight to recognize that something may happen in the future. These analysts work to anticipate the probability and impact of a threat and decide when to push action forward. Proactive action backed by effective cyber intelligence against a threat can stop a threat before it causes your organization harm. A cybersecurity analyst backed by intelligence can do wonders for improving an organization's security posture. As Sun Tzu mentioned in his book *The Art of War*, you must know your enemies well if you want to win the fight over them. Threat identification can be used to proactively determine the risk level of impact over the organizational assets and then help minimize the potential impact if the threat occurs in the organization's environment. For the Fusion Center to be an effective component of the organization's cyber defense program, it needs to gather and analyze data to determine whether the threat is applicable to their organization. From there, the Fusion Center should drive the proper level of remediation urgency. It is also important for Fusion Center cybersecurity analysts to recognize that the intelligence they are working with can also misguide them into pushing for the incorrect course of actions if they are using low-quality threat gathering feeds. There is a lot of misinformation available that can lead inexperienced analysts to incorrect conclusions that cause too much work for their organization. Often this additional work has little relevance to preventing true risk. Sometimes ambiguity can overshadow facts in the world of cybersecurity intelligence analysis. Things become hazy, speculation occurs, and the conclusions do not add up to your expectations. When this happens, it is important to take the time to locate and analyze more quality threat feeds to determine the proper risk impact of the event to your organization. Effective Fusion Center operatives understand that cyber intelligence risk assessment is about maintaining a balance between distraction and certainty, between calling wolf and protecting the organization. This work is both a profession and an art form. And when done correctly, it can lead to proactive organizational defense. Cybersecurity intelligence is complex and subject to interpretation. It is only as good as the source it is acquired from, the time it is acquired, and the expertise of the intelligence analyst.

Here are some key decisions to consider when deciding to build out SOC operations for your company: 24×7 operations, after-hours on-call operations, full-time employees (referred to throughout this book as FTEs), contractor operated, onshore, near-shore, offshore. Will you use an all-in-one managed service security provider, a hybrid, or will you equip it and run it? We will cover most of the items in this decision space in Chapter 2. There are two particularly important items to understand whether you decide to build up the SOFC in-house or with a managed security service (MSS) solution:

1. You should not think of such a critical asset as fire and forget. You must have an internal leader who has a dedicated focus on ensuring that the SOFC is working as intended and that the analysts are passionate about protecting your organizational data and assets. It is too easy to fall into the trap of thinking, "I bought an MSS, so I am fine and do not have to worry about doing anything else." It simply does not work that way. The MSS will find many items that they can handle, but there will be times when they need an internal resource who can drive resolution or provide oversight when a major event occurs. Even the best MSS providers can become complacent and lose their drive and passion if left unattended and uncared for.

2. SOFC operations are all about quality and not quantity or the pedigree of the provider. It is about having a team of

dedicated, focused, passionate individuals who are driven to protect your organization. The key is to exude dedication and commitment to protecting your organization. When using a pool of resources, it is critical that you know how many other companies they are having to watch and how much daily focus they are really giving to your organization. As Radziwill stated in his treatise on the quality of cost in cybersecurity, quality is a key and foundational element for achieving and maintaining competitiveness across the cybersecurity operational landscape.

As businesses continue to become more reliant on technology, the risk of cyber threats increases significantly. An SOC serves as a central hub for an organization's cybersecurity defenses. Its primary function is to continuously monitor and defend against cyber threats while also communicating awareness of these threats to key stakeholders. As previously mentioned, typically, the SOC is staffed with a team of cybersecurity experts with a range of skill sets. Together, this team works to protect an organization's data and assets from a broad range of threats such as malware, phishing attacks, hacking attempts, and other cyber threats. The primary goal of an SOC is to quickly detect and respond to potential threats to an organization's network and infrastructure. This is achieved through real-time monitoring of network activity, as well as proactive threat hunting and analysis. When a threat is detected, the SOC team will take immediate action to contain the threat and minimize any potential damage. In addition to actively defending against cyber threats, an SOC also plays a critical role in educating an organization's staff about cybersecurity best practices. This includes providing training on how to recognize and respond to potential threats, as well as promoting a culture of security awareness throughout the organization. When the threat intelligence center work is combined with an SOC to create a blended team that I am calling the SOFC it becomes a vital component of any organization's cybersecurity defense strategy. By leveraging the expertise of a team of dedicated security professionals, an organization can significantly reduce its risk of falling victim to a cyberattack.

2

Designing the SOFC Model

Even the most skilled warriors and cunning outlaws are not immune to mistakes, for even a ninja blinks and pirates are not perfect.

Dr. Kevin Lynn McLaughlin, PhD

When cybersecurity teams have initial discussions about what their Security Operations Center (SOC) or Cybersecurity Operations and Fusion Center (SOFC) will look and feel like, they should ask themselves if they should use a managed security service provider (MSSP) to build it out or whether they should use an in-house solution to build their own. This book will focus on building out an SOFC and not an SOC, though many of the concepts being discussed are relevant to both. The size of the organization and availability of sufficient talent to do the work, combined with the cost per year per person, are key factors in the decision to staff and build an SOFC internally, or to use an MSSP. A 24×7 employee staffed SOFC can be costly for many organizations. The minimum dollar amount to staff a 24×7 operation and allow for vacations, sick days, etc. can be more than $150,000.00 per year. And the reality is, an SOFC analyst with appropriate management oversight is going to cost more than that. Many companies begin to lose their enthusiasm for having a cybersecurity operations center once the cost of facilities, equipment, and gear is factored in – even though companies know they need to continuously monitor what is going on across their organizational landscape. Remember, an SOFC is both a team of people operating around the clock, and a facility dedicated to preventing, detecting, assessing, analyzing, and responding to cybersecurity threats and incidents. This being the case, a co-managed SOFC that uses an MSSP can be a viable option. If cost is one of the main factors for the set-up and deployment of your SOFC, you really need to consider using an MSSP in a low-cost region. If the

     DOI: 10.1201/9781003259152-3

Table 2.1   Examples of Key Consideration Criteria

CRITERIA	TYPE	WHAT IS ALLOWED
Size of organization		
Worth of organization		
The impact a negative cyber event would cause (money and revenue)		
Organization's operating model	In-house	US only
	Outsourced	Near shore
	Hybrid	Offshore
Budget		
Shared or dedicated analyst		

decision is made to utilize an MSSP, it is then important to determine if a pool of resources are preferred or if dedicated resource allocations would be acceptable. This part of the SOFC journey requires multiple components to be considered. Table 2.1 represents some of the key decision-making criteria that are used when deciding to internally staff an SOFC versus leveraging an MSS partner.

A key differentiator that determines whether an MSSP is required over an in-house SOFC is based on multiple factors. There is no formal or completely quantifiable gauge that can tell us to use an MSSP over an in-house SOFC. The decision model shown above can help in identifying whether your organization should leverage an MSSP, look to build the capabilities in-house, or stand up a hybrid SOFC solution. There is no silver bullet answer on this one; there are plenty of arguments that can be used to justify either approach. Look me up at a future security conference, and we'll go out for dinner and drinks and spend the night debating the merits of one over the other. Some of the biggest enterprises in the world have an in-house SOFC, and some of the biggest enterprises in the world leverage an MSSP to run their SOFC. This is discussed in more detail in Chapter 11. Below are some items to consider when making this decision:

1. Do you have a fixed budget for the creation of your SOFC? Budget is one of the more important variables. Depending on the MSSP you select, your analyst cost will vary between $20.00 per hour and $110.00+ per hour. Keep in mind that regardless of what the more costly MSSP providers say, you are not necessarily getting what you pay for. There are plenty of

lower cost MSSP solutions that you can put in place that will do an excellent job protecting your organization.

2. Analyst retention and focus. With an MSSP you do not have any recruiting worries. Retention becomes someone else's problem. A big difference between building out your own SOFC and utilizing an MSSP is that you do not have to worry about security talent acquisition. You will have specialized talent to back your security operations without an overhead to manage the team.

When considering the cost model of utilizing an MSSP versus an in-house solution for your SOFC, there are several advantages to be considered. An MSSP offers several benefits to organizations, including a predictable and fixed cost model. This makes budgeting and forecasting easier as the projected year-over-year cost for SOFC operations is clearly understood. Additionally, the MSSP handles the day-to-day management and coordination of these operations. This frees up the organization to focus on other important tasks. Furthermore, MSSPs can bring a wider range of expertise and specialized skillsets – offering access to the latest security technologies and best practices. These technologies and best practices can improve an organization's overall security posture. Lastly, by leveraging the MSSP's economies of scale, organizations can take advantage of resources and expertise that would otherwise be costly or unavailable. Using an MSSP can also provide your organization with greater flexibility, allowing you to scale your security operations up or down as needed. This can help ensure your organization is able to meet its security needs in an efficient and cost-effective manner. I cannot overemphasize the importance of carefully evaluating the specific needs and requirements of your organization before deciding on whether to utilize the services of an MSSP. It is crucial to weigh the potential benefits of a fixed cost model, specialized expertise, and economies of scale. And let's not forget the benefits of flexibility against potential drawbacks such as loss of control, lack of personalization, and dependence on an external provider. In the event of a prior or ongoing breach, having a strong and trusted MSSP with a reputable presence in the cybersecurity industry can be an advantage. Your partnership with an MSSP will enable you to leverage their expertise beyond that of the SOC analyst and leaders, which allows you to quickly bring in forensic experts, recovery specialists, and remediation specialists to help speed up recovery and rebuild your compromised IT infrastructure. For publicly traded companies, having an SOC or SOFC in place to monitor, detect, and respond to security alerts is considered a fiduciary responsibility. If your organization does not have the time or desire to build internal SOFC expertise, utilizing an MSSP to direct logs to their infrastructure and analyst pool for follow-up and remediation may be a practical solution. However, it is important to note that someone within the organization with the necessary bandwidth and expertise must be available to properly triage, review, and manage the alerts received. For organizations facing both cost constraints and challenges in acquiring and retaining necessary security talent, utilizing an MSSP can provide access to a larger pool of security talent and prove to be a more cost-effective solution. This is because having an MSSP eliminates the need to invest in an expensive security infrastructure. There are other factors that you need to consider when deciding whether to choose an in-house security solution or an MSSP. The size of your security staff and their capability, specialization, and skillsets play a crucial role in this decision. If your organization requires a large security team with specialized skillsets, an MSSP may be a better option, as they have a larger pool of resources available. On the other hand, if your organization is small, an in-house solution may be more cost-effective and efficient.

Another factor that you should consider is whether you have a chief information security officer (CISO) or are planning to hire one. If you have a CISO in place, an in-house solution may be more appropriate, as your in-house team can work closely with the CISO to ensure alignment with your organization's overall security strategy. However, if you do not have a CISO, an MSSP may be a better option, as they can provide the necessary guidance and expertise. The presence of a chief information officer (CIO) in your organization is another item that should be taken into consideration. If you have a CIO in place, an in-house solution may be more appropriate, as they can work closely with the CIO to ensure alignment with your organization's overall technology strategy; if you do not have a CIO, an MSSP may be a better option, as they can provide the necessary guidance and expertise.

The legal and governance guidelines within your organization are other important factors to consider. If your organization has clear legal guidelines in place, an in-house solution may be more appropriate, as it allows for better alignment with your organization's legal/compliance requirements. However, if your organization does not have clear legal guidelines, an MSSP can provide the necessary guidance and expertise. If your organization has a clear cybersecurity governance directive in place, an in-house solution may be more appropriate, as it allows for better alignment with your organization's overall security strategy. If your organization does not have a clear security governance directive, an MSSP can provide the necessary guidance and expertise. Finally, the current security status of your organization needs to be considered. If your organization is currently under attack, an MSSP may be a better option, as they can provide the necessary resources and expertise to quickly respond to an incident. On the other hand, if your organization is currently cybersecure, an in-house solution may be more appropriate, as it allows for better alignment with your organization's overall security strategy.

When opting for an MSSP to handle your security operations, it is important to note that you will still need a dedicated team within your organization to act as the liaison between the MSSP and the various technology teams within your company. This team should consist of two to three individuals to ensure effective communication and coordination. It is important to note that an MSSP will typically have limited access and scope when it comes to communication with the broader organization. While this may seem disadvantageous, it can be beneficial in preventing unnecessary panic and disruptions to business operations.

In the field of cybersecurity, it is essential to maintain a level of discretion and confidentiality. Security operations can spark curiosity and interest among employees that can potentially lead to speculation and gossip. By limiting the scope and group of individuals within the organization who have access to security information, the focus can remain on addressing and resolving security threats, as opposed to addressing speculation and conspiracy theories. This approach can be likened to a skilled ninja wielding a sharp sword who is able to navigate through obstacles efficiently and effectively, rather than being hindered by them. Cost comparison is an important consideration when choosing between utilizing an MSSP for your SOFC versus an in-house solution. When comparing the cost of an MSSP SOFC to an in-house SOFC, it is important to consider the total cost of ownership (TCO) over a certain period. Usually, you want to consider these costs on an annual basis. The TCO of an MSSP SOFC includes the cost of the services provided by the MSSP, as well as any additional costs associated with managing and maintaining the SOFC. These additional associated costs include hardware, software, and personnel costs. On the other hand, the TCO of an in-house SOFC includes the cost of personnel, hardware, software, and any additional costs associated with managing and maintaining the SOFC. It also includes the cost of recruiting, hiring, and training security personnel, as well as the cost of any necessary upgrades or maintenance for the hardware and software used by the SOFC. When comparing the TCO of an MSSP SOFC to an in-house SOFC, it is important to note that an MSSP SOFC can have a fixed cost model, which allows for more accurate budgeting and forecasting. This can be beneficial for organizations that have limited resources and need to carefully manage their expenses. However, an in-house SOFC can provide greater flexibility and customization, as the organization has more control over the personnel, hardware, and software used in the SOFC. Additionally, an in-house SOFC may be more cost-effective in the long term, as the organization will not have to pay for the ongoing services provided by the MSSP. It is important to carefully consider the specific needs and requirements of your organization before deciding and weighing the benefits and drawbacks of MSSP and in-house SOFC solutions. The decision should be based on a thorough cost-benefit analysis, considering the specific needs, resources, and risk appetite of the organization.

An advantage of using an MSSP is the access to a wide range of security experts and resources. MSSPs employ specialized security personnel with a broad range of skills and expertise, which provides organizations with access to a vast amount of knowledge and experience. This can be especially beneficial for small and medium-sized organizations that lack the resources to build and maintain an in-house SOFC and Fusion Center. On the other hand, one of the main disadvantages of using an MSSP is the lack of control and customization. Organizations that use MSSPs may have limited control over the security services provided and may not be able to fully customize the solution to meet their specific needs. Additionally, MSSPs may not have open communication with the broader organization and may have a limited security scope. An in-house solution for your SOFC and Fusion Center can provide greater flexibility and customization, as the organization has more control over the personnel, hardware, and software used in the SOFC. Additionally, an in-house SOFC may be more cost-effective in the long term, as the organization will not have to pay for the ongoing services provided by the MSSP.

I understand we went through a lot of back and forth in this chapter; future chapters will focus and clarify the in-house versus MSSP discussion. One of the most critical decisions an organization must make when it comes to cybersecurity defense is whether to build and staff an internal cybersecurity SOFC or to use a co-managed SOFC that partners with an MSSP. The decision to go one way or the other depends on a range of factors, such as:

- the size of the organization
- the availability of cybersecurity talent
- the cost per year per person
- the depth of cybersecurity staff already working in the organization
- how much flexibility is needed for day-to-day process

For larger organizations with a significant budget, building and staffing an internal SOFC can make sense. Doing so enables the organization to have direct control over its cybersecurity defense strategy and ensures that the team is familiar with the company's network and infrastructure. With a dedicated team of experts, an internal SOFC can also provide a more in-depth level of protection against a wide range of cyber threats. However, smaller organizations may struggle to find the necessary talent to staff an internal SOFC, and the cost per year per person may be prohibitive. In these cases, a co-managed SOFC in a low-cost region that partners with an MSSP can be a viable option. This approach allows the organization to leverage the expertise and resources of a third-party provider to help monitor and defend against cyber threats. A co-managed SOFC can be an especially attractive option for smaller organizations that are just beginning to develop their cybersecurity defense strategy. In the end, there are plenty of arguments that can be made for either approach. Building an internal SOFC provides direct control and an in-depth level of protection but can be expensive and difficult to staff. A co-managed SOFC with an MSSP, on the other hand, can be a more cost-effective option, but may not offer the same level of control or flexibility. The decision comes down to what makes the most sense for the organization based on its unique circumstances, priorities, and budget. Regardless of which method an organization goes with they need to have someone on point for training the SOFC on internal process, procedures, and culture if they want their SOFC to be successful. Even when you hire the best-in-class MSSPs the organization still needs to provide training, guidance, and oversight to ensure incidents and threats are handled correctly and expedited with the sense of urgency the internal cybersecurity team expects.

3
SOFC
Building the Core Team

An effective and high-performing Security Operations and Fusion Center (SOFC) is the cornerstone of a top-notch cybersecurity program.

Dr. Kevin Lynn McLaughlin, PhD

Building the SOFC Core Team

It is important to put together a highly skilled and knowledgeable core team to establish an effective and efficient Cybersecurity Operations and Fusion Center (SOFC) for your organization. Building a top performing SOFC team involves a comprehensive approach that includes recruitment, retention, and ongoing training and development. The recruitment process is a critical aspect of building an SOFC team. It is important to take your time to identify and attract individuals who possess the knowledge, skills, and experience needed to drive success. This will help ensure that your SOFC team has the capabilities required to effectively monitor, detect, and respond to security alerts, and to continuously improve the overall security posture of your organization. Retention is another key aspect of building a successful SOFC team. You must have strategies in place to retain top talent and ensure that your SOFC team stays engaged and motivated. This can include regular training and development opportunities, competitive compensation packages, and a positive and supportive work environment. Ongoing training and development is also crucial for the success of your SOFC team. As stated in previous ISACA workplace studies on cybersecurity training, development and training is sometimes more important to your team member than salary. The cybersecurity landscape is constantly evolving, so it is essential to ensure that your

DOI: 10.1201/9781003259152-4

SOFC team members are kept up-to-date with the latest techniques, technologies, and best practices. This will help your team provide the highest level of service to your organization while continuously improving its overall security posture.

Choosing Your Analyst

The first step in building your SOFC team is selecting the right analysts. The ideal candidate should possess a strong understanding of the industry and the specific business processes and systems the SOFC will focus on. It is my belief that an effective SOFC cybersecurity analyst should possess a unique set of traits that enables them to effectively detect, respond, and mitigate cyber threats. Primarily, the analyst should possess a strong technical aptitude and be versed in the various technologies and systems used in the industry. This includes an understanding of operating systems, networks, and security protocols. Additionally, the analyst should have an understanding of the various threat actors and their tactics, techniques, and procedures. In addition to technical skills, an effective cybersecurity analyst should possess strong analytical and critical thinking abilities. The ability to analyze substantial amounts of data, identify patterns and anomalies, and draw meaningful conclusions is crucial for detecting and responding to cyber threats. The ideal analyst should also have effective communication skills. In today's fast-paced and highly connected world, it is essential for analysts to be able to communicate effectively with a diverse group of stakeholders, including IT teams, management, and other security professionals. The ability to explain technical concepts in a clear and concise manner is crucial in facilitating effective incident response and communication. Furthermore, an effective SOFC cybersecurity analyst should be a continuous learner. The cybersecurity landscape is constantly evolving, with new threats emerging every day. A successful analyst should be able to stay current on the latest developments in the field and be willing to continuously learn new skills and technologies.

Lastly, an effective analyst should have an intense sense of ethics and integrity. The protection of sensitive data and the safeguarding of an organization's assets is a critical responsibility. It requires an intense sense of ethics and integrity. An analyst should be committed to upholding the highest standards of professional conduct while protecting an organization's assets and reputation. In short, an effective cybersecurity analyst should possess a unique combination of technical skills, analytical abilities, effective communication skills, a continuous learning mindset, and a powerful sense of ethics and integrity. These traits enable them to effectively detect, respond, and mitigate cyber threats and protect an organization's assets and reputation.

Choosing Your Senior Analysts

In addition to cybersecurity analysts, it is also important to have senior-level SOFC cybersecurity analysts on the team who can serve as leaders and mentors for the rest of the group. Senior analysts should have a proven record of success in their roles and possess a deep understanding of the industry and business processes. They should also be able to lead and manage teams effectively, as well as possess strong strategic thinking and problem-solving skills. There are several key traits that are essential to the success of a senior cybersecurity analyst within the SOFC. Primarily, a senior SOFC analyst must possess a deep understanding of cybersecurity concepts and technologies. This includes knowledge of diverse types of cyber threats, as well as an understanding of security best practices and industry standards. Additionally, the senior analyst should have a thorough understanding of the various tools and technologies used in the SOFC, such as intrusion detection and prevention systems, firewalls, and security information and event management systems. In addition to technical expertise, a senior SOFC analyst must possess strong analytical and problem-solving skills. This includes the ability to quickly and accurately identify and triage security incidents, as well as the ability to evaluate and prioritize risks to the organization. The senior analyst should also be able to effectively communicate findings and recommendations to both technical and non-technical stakeholders. Other crucial traits of a senior SOFC analyst include effective communication and teamwork. The senior analyst must be able to work well with others, both within the SOFC and cross-functionally with other teams inside the organization. This includes the ability to collaborate on incident response and threat hunting efforts, as well as the ability to effectively communicate with incident response teams, IT, and business stakeholders.

Finally, a senior SOFC analyst should possess a keen sense of curiosity, and a desire to stay current with the latest cybersecurity trends, threats, and technologies. This includes a willingness to continuously learn and develop new skills, as well as a desire to share knowledge and mentor junior analysts within the team. In summary, an effective senior SOFC analyst should possess a deep understanding of cybersecurity concepts and technologies, strong analytical and problem-solving skills, effective communication and teamwork, and a powerful sense of curiosity and desire to stay current with the latest trends and technologies. These traits will enable the senior analyst to effectively protect the organization from cyber threats and respond to security incidents in a timely and efficient manner.

Recruitment Process

The recruitment process is an essential element of building a successful SOFC team. It is important to identify the right candidates by using a thorough and comprehensive process that includes resumes, interviews, and reference checks. You should appeal to candidates by highlighting the exciting opportunities and challenges that the SOFC role offers, as well as the benefits of working in a dynamic and innovative environment. As a cybersecurity practitioner with a PhD in the field and over three decades of experience, I can attest to the importance of effectively recruiting top talent for an SOFC team. Recruiting the right individuals for the role of SOFC analyst and senior analyst is crucial for the success and effectiveness of the team. When recruiting SOFC analysts and senior analysts, it is important to look for individuals who possess a strong technical background in cybersecurity. This includes knowledge of various security technologies and methodologies, as well as experience with incident response, threat hunting, and security analysis. In addition to technical skills, it is also important to look for individuals who possess strong analytical and critical thinking abilities. SOFC analysts and senior analysts must be able to quickly process and analyze copious amounts of data, identify patterns and anomalies, and make informed decisions based on that data. Another important trait to look for in SOFC analysts and senior analysts is effective communication skills. These individuals must be able to effectively communicate with other members of the SOFC team, as well as with other departments within the organization. They must also be able to explain complex technical concepts clearly and concisely to non-technical stakeholders. When it comes to the recruitment process, companies can conduct technical interviews, skills assessments, and background checks to evaluate the candidate's skills, knowledge, and qualifications. Experience in similar roles, certifications, and education in the field of cybersecurity can also be taken into consideration. It is also crucial to consider the cultural fit of the candidate within the organization. A candidate who aligns with the company's values and culture is more likely to be a long-term asset to the team and remain invested in the success of the organization. Effective recruitment for SOFC analysts and senior analysts should focus on identifying individuals who possess strong technical skills, analytical abilities, and communication skills while being a strong cultural fit.

Appealing to Your Prospects

Appealing to potential candidates is key to building a successful SOFC team. The recruitment process must be designed to highlight the organization as a top-performing and innovative company with a strong focus on teamwork and collaboration. Highlighting the company culture, the opportunities for growth and career advancement, and the benefits and rewards of working in the organization can help attract top talent. As a leader in the field of cybersecurity, it is important to understand the expectations and needs of the cybersecurity SOFC analysts and senior analysts working under your guidance. These professionals are highly skilled and educated in the field – this comes with certain expectations of their leaders. Analysts and senior analysts are looking for clear and consistent communication and direction from their leaders. This includes being kept informed of any changes to the organization's security posture, as well as any updates to policies, procedures, and protocols. Cybersecurity professionals also expect their leaders to be accessible and responsive to their needs and be willing to listen to their feedback and suggestions. In addition to clear communication, analysts and senior analysts are looking for leaders who can provide them with the necessary tools and resources to do their job effectively. This includes access to the latest technologies, training, and development opportunities. Another important aspect of effective leadership in the cybersecurity field is the ability to foster a culture of collaboration and teamwork. This includes creating a work environment that encourages the sharing of ideas and the pooling of knowledge and skills. Leaders are expected to be supportive of their team's professional growth and development. I often tell my team: we win together and when we fail, we fail together. Helping each other is not a sign of failure or ineptness, but failing to work together to be successful is. Effective recruitment for cybersecurity SOFC analysts and senior analysts also includes looking for individuals who possess a strong work ethic. That work ethic includes a willingness to take on challenging tasks and a desire to stay current with the latest trends and developments in the field. The ability to think critically and analytically with the use of strong problem-solving skills are other essential traits for both analysts and senior-level analysts. To sum it up, effective recruitment for cybersecurity SOFC analysts and senior analysts involves looking for individuals who possess the necessary technical skills and knowledge, as well as the ability to work well with others. It's also important to ensure your recruits have a desire to stay current with the latest trends and developments in the field. As leaders, you should be able to provide your team with the necessary tools and resources to do their jobs effectively, foster a culture of collaboration and teamwork, and be supportive of their professional growth and development.

Team Relationship Building

As a leader in the field of cybersecurity, it is important to recognize the unique challenges that come with managing a team of SOFC analysts and senior analysts. These individuals are tasked with the critical responsibility of identifying and mitigating cyber threats daily. This can be a stressful and demanding job. One of the key ways that leaders can support their team is to focus on building an intense sense of camaraderie and teamwork. This can be achieved through a variety of team-building initiatives and activities, such as regular team meetings, team-building exercises, and opportunities for team members to bond and connect outside of work. Another important aspect of leading a successful SOFC team is to provide clear and effective communication. This includes ensuring that team members are aware of their roles and responsibilities, as well as providing them with the training and resources they need to effectively perform their duties. Leaders should strive to create an environment that fosters a sense of ownership and autonomy among team members. This can be achieved by providing them with the tools and resources they need to do their job effectively, and by giving them the freedom to innovate and develop new strategies for identifying and mitigating cyber threats. Finally, it is important for leaders to lead by example. Leaders can do this by demonstrating a commitment to the team's success through their own actions. This means being available to team members for support and guidance, and providing the team with the resources they need to do their job effectively.

Overall, effective recruitment and team building initiatives are crucial for any cybersecurity SOFC analyst and senior analyst to be successful in their role. It is important for leaders to support and empower their team members, and to foster a positive work environment that encourages collaboration, innovation, and teamwork. Once the team is assembled, it is essential to focus on building strong relationships within the group. This can be achieved through regular team-building activities, such as team lunches, off-site retreats, and team-building exercises. Additionally, it is important to promote a diverse, open, and inclusive environment where all team members feel comfortable sharing their ideas and opinions.

Retention Strategies

Retention is a key component of building a successful SOFC team. To keep team members engaged and motivated, it is important to implement retention strategies like gamification, which can help increase team pride and passion for the company. It is also imperative to provide opportunities for ongoing training and development to help team members grow and advance in their roles. As a senior cybersecurity expert, it is important to have a comprehensive understanding of the key retention strategies for Cybersecurity SOFC analysts and senior analysts. Here are a few tips to consider:

- **Provide Opportunities for Growth and Development:** Offering your analysts and senior analysts training, certifications, and advancement within the organization can help keep your team engaged and motivated.

- **Build a Positive and Inclusive Work Culture:** Creating a positive and inclusive work culture where analysts and senior analysts feel valued and respected is essential for retention. This includes fostering a sense of belonging and encouraging open communication.
- **Recognize and Reward Excellent Work:** Recognizing and rewarding excellent work is essential for employee engagement and motivation. This can be done through performance bonuses, promotions, and public recognition.
- **Provide a Competitive Compensation Package:** Offering a competitive compensation package, including salary, benefits, and rewards, can help attract and retain top talent in the field.
- **Prioritize Work-Life Balance:** Prioritizing work-life balance is important for the well-being of analysts and senior analysts. This includes offering flexible working hours, remote work options, and opportunities for paid time off.
- **Provide Regular Feedback and Coaching:** Regular feedback and coaching can help analysts and senior analysts understand their strengths and weaknesses, and develop the skills needed to excel in their roles.
- **Encourage Collaboration and Teamwork:** Encouraging collaboration and teamwork among the cybersecurity SOFC analysts and senior analysts not only improves the team's performance, but also helps create a sense of camaraderie and belonging among the team members.

Effective retention strategies for cybersecurity SOFC analysts and senior analysts involves creating a positive work environment, providing opportunities for growth and development, and recognizing and rewarding excellent work.

Ongoing Training, Growth, and Exercises

To ensure that the SOFC team can achieve success in the long-term, it is essential to provide ongoing training and development opportunities. This can include workshops, webinars, training sessions that focus on the latest industry trends and best practices, formal classes as well as informal learning opportunities, and skill building such as mentoring and job shadowing. Regular team exercises and simulations can help keep the team sharp and prepared for any challenges that may arise. Providing access to online resources, such as webinars and online training programs, is an effective way to ensure that analysts have the knowledge they need to stay up-to-date with the latest developments in the field. It's also important to provide opportunities for advancement and career development. This can include promoting from within, offering leadership and management training, and providing opportunities for analysts to work on special projects or to take on additional responsibilities. You can even look at providing incentives for analysts to further their education, such as tuition reimbursement or professional certification programs. This can be an effective way to retain top talent and attract new hires. It is essential to foster a culture of continuous learning and improvement within the cybersecurity team. This can include setting clear performance expectations and regularly providing feedback, as well as recognizing and rewarding analysts for their contributions to the team. You're going to want to provide opportunities for team members to share their knowledge and expertise with one another through regular team meetings and even knowledge-sharing sessions. This can help build a culture of collaboration and learning. Overall, effective training and professional growth strategies for cybersecurity SOFC analysts and senior analysts are essential for ensuring these individuals have the skills and knowledge they need to protect organizations from cyber threats. By providing regular education and training opportunities, promoting from within, and fostering a culture of continuous learning and improvement, organizations can retain top talent and attract new hires. This also helps ensure your cybersecurity teams are equipped to meet the challenges of the constantly evolving threat landscape.

Teaching the 4 S's

One of the key elements of building a high-performing SOFC team is teaching the 4 S's: *Seek to Understand, Share Ideas, Strength in Numbers,* and *Study.* These principles focus on the importance of effective communication, teamwork, and continuous learning and development. By emphasizing these principles and building them into the team culture, organizations can create a cohesive and effective team that can achieve remarkable success. As a seasoned cybersecurity professional and researcher, I have found that the key to building a highly skilled and effective team within your SOFC is to instill a culture of continuous learning and development. One way to do this is through the teaching of the 4 S's.

The first S, "Seek to Understand," encourages team members to actively listen and learn from one another, to understand different perspectives, and to be open to new concepts and ideas. This helps create a culture of collaboration and mutual respect within the team. The second S, "Share Ideas," encourages team members to share their knowledge and expertise with one another, to freely exchange ideas, and to be open to constructive criticism. This helps foster a culture of innovation and creativity within the team. The third S, "Strength in Numbers," emphasizes the importance of teamwork and collective efforts in achieving success. This helps create a sense of shared responsibility and accountability within the team and encourages team members to work together to achieve common goals. The fourth S, "Study," urges team members to continuously learn and develop their technical and soft skills. This can include attending training and certification programs, reading industry publications, participating in research projects, and attending conferences. This helps create a culture of continuous learning and development within the team and ensures that the team members are up-to-date with the latest developments in the field. By teaching the 4 S's, leaders can help build a high-performing SOFC team that is well-equipped to handle the challenges of the ever-evolving cybersecurity landscape.

Gamification for the Win

A core component of training, recruiting, enticing, retaining, and creating a high performing, highly engaged SOFC workforce is to leverage the power of gamification. As I once stated at an RSA conference I was lucky enough to speak at,

> Gamification is a crucial tool for any organization looking to attract, retain, and motivate top cybersecurity talent. By incorporating elements of competition and reward, gamification can

help to create an engaging and dynamic work environment that keeps employees motivated and invested in their work.

I have seen and experienced the impact that gamification can have on the recruitment, enticement, retention, and motivation of cybersecurity professionals first-hand. From wargames that simulate real-world cyberattacks, to capture the flag (CTF) competitions that test the skills and knowledge of participants, to cyber ranges that provide a safe and controlled environment for hands-on experience – the possibilities are endless. These gamification strategies not only engage and challenge cybersecurity talent, but they also provide valuable training and development opportunities that can help build a stronger, more resilient workforce. When you use gamification it's important to think of rewards that resonate with the cybersecurity professionals you hire. Consider items such as Ninja ranks, achievement patches, certificates that name the greatest of the great throughout the land, and other items of interest. I have provided threat hunters with an authentic air force bomber jacket, complete with custom patches after they've captured five bad actors who were lurking in our environment. I was amazed at the amount of work, attention, and effort the threat hunters would exude to receive one of those jackets. Here are some examples of rewards to consider:

- Access to exclusive cyber ranges and challenges, such as simulated attack and defense scenarios.
- Opportunities to participate in prestigious cybersecurity competitions and events, such as CTF or wargames.
- Recognition and bragging rights for exceptional performance and achievements within the community, such as winning awards or being listed on leaderboards.
- Opportunities to continuously learn and grow their skills through customized training programs and resources.
- Access to unique and interesting projects and initiatives that align with their personal and professional goals.
- A set of ninja-themed patches, representing different levels of achievement in a particular area of expertise.
- A limited edition, military-style jacket, exclusively for top-performing members of the team.

- A certificate of recognition, signed by a renowned security expert, for exceptional performance in a wargame or CTF competition.
- An invite to an exclusive, invitation-only gathering of cybersecurity legends and thought leaders, for the elite few who have made a significant impact in their field.

This chapter was specifically designed for leaders in the cybersecurity field who are seeking to create a dynamic and cohesive SOFC team of analysts and senior analysts. We discussed the critical importance of selecting the right people for the team and detailed insightful guidance on the recruitment process. It's important to appeal to potential team members and to cultivate positive relationships within the team. Strategies for retaining top talent within the SOFC were also shared, such as incorporating gamification and offering ongoing training and growth opportunities. We explored the teaching of the 4 S's: *Seek to Understand, Share Ideas, Strength in Numbers*, and *Study*, and determined that they are essential for fostering a strong team culture and cohesion. By following the guidelines and strategies outlined in this chapter, you will be able to build and maintain a high-performing team within the SOFC. Hopefully, this chapter provided valuable information that can help you create a positive and productive work environment, allowing your organization to achieve its goals. By reading this chapter, you will be able to take your understanding of building a strong team within the SOFC to the next level. The information and strategies presented will help empower you to create a dynamic, cohesive, and high-performing team that will drive your organization to success.

PART II
TOOLS AND OPERATIONS

4
SOFC INFRASTRUCTURE AND TOOLSET

As a cybersecurity expert, it's imperative that one's toolkit includes AI and ML technical solutions to stay ahead in the ever-evolving cyber combat battlefield.

Dr. Kevin Lynn McLaughlin, PhD

In this chapter, we'll delve into the important topic of designing the Cybersecurity Operations and Fusion Center (SOFC) infrastructure and toolset. The infrastructure and toolset are crucial for the success of the SOFC. They provide the foundation for detecting and monitoring activity across the network. A cloud-first strategy is critical to the success of an SOFC for several reasons. First, a cloud-first strategy enables organizations to take advantage of the scalability and flexibility of the cloud. This means, as the organization's security needs change and evolve, the SOFC can easily scale up or down to meet those needs. This is particularly important in today's fast-paced and ever-changing cybersecurity landscape, where new threats are constantly emerging and existing threats are frequently evolving. A cloud-first strategy also allows organizations to take advantage of the latest and most advanced security technologies. Many of the top security vendors are now offering their products and services in the cloud, which means that organizations can take advantage of these innovative technologies without having to invest in an expensive on-premises infrastructure. Another key element of a cloud-first strategy is that it enables organizations to share information more effectively and collaborate with other organizations. By leveraging the cloud, organizations can easily share threat intelligence and other security-related information with other organizations. This can help organizations better understand and respond to threats. Finally, a cloud-first strategy allows organizations to comply with regulations such as SOC2, PCI-DSS,

DOI: 10.1201/9781003259152-6

and HIPAA more easily. Meeting these regulatory requirements helps organizations avoid costly penalties and fines. To sum it up, a cloud-first strategy is critical to the success of a cybersecurity SOFC because it allows organizations to take advantage of the scalability, flexibility, and advanced technologies of the cloud, while also allowing them to share information more effectively, collaborate with other organizations, and comply with regulations and standards.

Physical Security of the SOFC

An SOFC must have proper and effective physical security controls in place to ensure that the center remains mission capable. This typically includes a secure location, with limited access and a comprehensive access control system that employs multiple layers of security. These security layers include biometrics, smart cards, and video surveillance. Physical security processes and procedures should be in place to ensure that all personnel and visitors are thoroughly vetted before they are granted access to the SOFC. This may include background checks, security clearances, and mandatory security training. Physical security measures for the SOFC are also necessary to protect the critical IT infrastructure within the SOFC, including servers, network equipment, and data storage devices. Locking up the rooms and cabinets where these devices are stored is essential. It's also important to implement physical security controls for the power and data connections to these devices. Finally, procedures should be in place for responding to physical security incidents, including intrusion alarms, power outages, and other security breaches. These procedures should be well documented and rehearsed, so the response team is able to quickly and effectively address any security incidents that may occur. Keep in mind that the physical security infrastructure, process, and procedures for protecting cybersecurity SOFC are critical to ensuring the integrity and confidentiality of sensitive data and information. Organizations can reduce the risk of an SOFC breach and maintain the confidentiality and availability of their critical information by implementing multiple layers of security and a comprehensive access control system. It's also essential to have well-defined procedures in place for responding to physical security incidents.

The SOFC Workforce

When it comes to the workforce, there are three main options: brick and mortar, hybrid, or remote. A remote workforce can offer several advantages. Allowing employees to work from home or other remote locations gives organizations the opportunity to tap into a larger pool of talent. This can be particularly useful when it comes to recruiting and retaining top cybersecurity professionals. Additionally, a remote workforce can lead to cost savings for the organization, as it reduces the need for expensive office space and other on-premises infrastructure. On the other hand, a hybrid workforce, which combines remote and onsite capabilities can provide the best of both worlds. It allows organizations to take advantage of the benefits of a remote workforce while also maintaining the benefits of onsite employees. This is particularly useful when it comes to fostering collaboration and teamwork, as well as providing access to onsite resources and infrastructure. An onsite workforce offers several advantages as well. By having employees work onsite, organizations can maintain tight control over their security operations and ensure they are always able to respond quickly and effectively to security incidents. Additionally, an onsite workforce can make it easier to comply with regulatory and compliance requirements, as well as foster a sense of teamwork and camaraderie among employees. Overall, the choice between using a remote workforce, hybrid workforce, or onsite workforce for SOFC cybersecurity analysts and senior analysts will depend on the specific needs and requirements of the organization. Organizations must evaluate the pros and cons of each option and make a decision that aligns with their overall security and business objectives.

It's important to note that remote and hybrid workforce models for SOFCs have become increasingly prevalent in recent years – particularly due to the COVID-19 pandemic. This new normal has forced organizations to rethink their approach to remote workforce operations and has highlighted the importance of proper standard operating procedures (SOPs). Leveraging virtual tools and technologies is a key aspect of a remote workforce. These tools allow for seamless collaboration and communication among team members, regardless of their physical location. The COVID-19 pandemic has led to a significant shift in the way cybersecurity professionals work. With many organizations turning to remote and hybrid working models, several productivity tools such as always on VPN connections, cloud-based end detection/ protection technologies and even items as simple as secure video conferencing technologies have become the new norm. One of the most important tools for remote and hybrid cybersecurity workers is video conferencing software. With many employees working from home, video conferencing has become a vital tool for communication and collaboration. Platforms such as Zoom, Microsoft Teams, and Google Meet have become increasingly popular for virtual meetings, team collaboration, and training sessions. Another useful tool for remote and hybrid cybersecurity workers is project management software. With many employees working remotely, it can be challenging for managers to keep track of the progress of different projects. Project management software such as Asana, Trello, and Monday have become increasingly popular for managing tasks, tracking progress, and keeping teams organized. Cloud-based storage and file sharing platforms have also become important tools for remote and hybrid cybersecurity workers. Platforms such as Dropbox, Google Drive, and OneDrive have become increasingly popular for storing and sharing files. This allows employees to access important documents and information from anywhere, at any time – which is crucial for remote and hybrid workers. Finally, secure remote access solutions, like virtual private networks and remote desktop protocols, have become essential tools for remote cybersecurity workers. These solutions allow employees to access their organization's network, resources, and applications securely. Overall, the COVID-19 pandemic has led to a significant shift in the way cybersecurity professionals do their jobs. As previously stated, several productivity tools have become the new norm for these hybrid and remote workers. These tools are critical for maintaining communication, collaboration, and productivity.

Standard Operating Procedures (SOPs)

SOPs provide a clear and consistent framework for the SOFC to follow in order carry out tasks and respond to security incidents. They outline the specific steps to follow to achieve desired outcomes. One example of a relevant SOP for an SOFC is an incident response plan. This plan outlines the steps to be taken in the event of a security incident. These steps include how to identify and contain the incident, how to gather evidence, and how to communicate with stakeholders. This SOP will provide the team with a clear and consistent approach to incident response, ensuring that all incidents are handled efficiently and effectively.

Another example of a relevant SOP is a security monitoring plan. This plan outlines the procedures for monitoring and analyzing security data. These procedures include what data to collect, how to collect it, and how to analyze it. This SOP will provide the team with a clear and consistent approach to security monitoring, ensuring that all potential threats are identified and addressed in a timely manner. An SOFC should also have an SOP for incident reporting, which should include the steps for reporting incidents, the format for reporting incidents, and the incident classification scheme. This will ensure incidents are reported consistently and in a way that is useful for further analysis. In conclusion, SOPs are critical for the successful operation of an SOFC. They provide a clear and consistent framework for carrying out tasks and responding to security incidents. They ensure that all team members are aware of their roles and responsibilities, and that appropriate actions are taken in a timely and efficient manner. By implementing SOPs, an SOFC can improve its overall effectiveness and efficiency, and ensure that it is better prepared to respond to and mitigate security incidents.

In this chapter we focused on the importance of designing the infrastructure and toolset for an SOFC. The chapter begins by highlighting the importance of a cloud-first strategy, which allows for greater flexibility and scalability when managing the SOFC. We then discussed brick and mortar, remote, and hybrid workforce options for the SOFC, with a focus on leveraging virtual tools and technologies to support a remote workforce. We also discussed the physical security of the SOFC. Physical security is crucial to ensure the protection of sensitive information and equipment. Towards the end of this chapter, we touched on the impact of COVID-19 on the SOFC workforce, and how it has shifted the thought process around the best approach for managing a remote workforce. We closed out this chapter by discussing the use of SOPs and their importance to successful SOFCs. SOPs provide clear and detailed instructions for how to carry out specific tasks. These instructions help minimize the risk of human error and ensure consistent results. Examples of relevant SOPs for an SOFC include incident response and security monitoring plans. Overall, this chapter provides readers with a comprehensive understanding of the various components and tools necessary for designing a successful SOFC infrastructure and toolset.

5

CYBERSECURITY OPERATIONS AND FUSION CENTER

Daily Operations

To triumph in the cyber realm against our foes, it requires a combination of fervor, mastery, and a steadfast determination to continuously learn and improve.

Dr. Kevin Lynn McLaughlin, PhD

In Chapter 5, we will look at the daily operations of the Cybersecurity Operations and Fusion Center (SOFC), including the duties and responsibilities of the analysts and team leaders. The SOFC is a critical component of an organization's cyber defense strategy.

The day-to-day activities of an SOFC analyst, senior SOFC analyst, and cyber intelligent analyst are fun, exciting, and constantly changing. Even those with eyes on glass (continuously monitoring a system from a computer) responsibilities find that their day is flexible while constantly thinking on their feet to stay ahead of threat actors.

A typical day for an SOFC analyst may involve the following activities:

- Monitoring and analyzing network traffic, endpoint protection software, and cloud services for potential threats and incidents.
- Reviewing and triaging alerts generated by security tools like Security Information and Event Management, endpoint detection and response, intrusion detection systems, and firewalls.
- Investigating potential incidents, collecting data, and determining the appropriate course of action.

DOI: 10.1201/9781003259152-7

- Communicating with other SOFC team members, incident response teams, and stakeholders to ensure effective incident response and resolution.
- Documenting incidents and their resolution for later analysis and improvement of processes and procedures.

A typical day for a senior SOFC analyst may involve the following activities:

- Overseeing the day-to-day operations of the SOFC, including monitoring, and managing the work of junior analysts.
- Conducting high-level investigations of complex and sophisticated incidents through the use of advanced analytical techniques and expertise.
- Developing and implementing new tools, processes, and procedures to improve the efficiency and effectiveness of the SOFC.
- Mentoring and training junior analysts to enhance their skills and knowledge.
- Collaborating with other teams, such as threat intelligence, incident response, and research, to ensure effective response and resolution of incidents.
- Briefing senior leaders on the outcome of incident response efforts and providing an overview and understanding of the SOFC metrics.
- Ensuring through quality control procedures that alerts are being handled as expected.

A typical day for an SOFC cybersecurity intelligence analyst may include the following activities:

- Monitoring open-source and other intelligence sources for potential threats, such as new malware variants, exploits, and hacking groups.
- Analyzing intelligence data to determine its relevance and credibility and identify potential threats.
- Collaborating with other teams, such as incident response and research, to ensure that relevant intelligence is shared and acted upon.

- Communicating with stakeholders, including internal customers and external partners, to share intelligence and provide guidance on risk mitigation.
- Documenting and tracking intelligence data to ensure that it is properly archived and available for future reference.

A typical day for an SOFC cybersecurity senior intelligence analyst may include the following activities:

- Overseeing the day-to-day operations of the intelligence team, including monitoring, and managing the work of junior analysts.
- Conducting high-level analysis of complex intelligence data through the use of advanced analytical techniques and expertise.
- Developing and implementing new intelligence-gathering and analysis processes and procedures to improve the efficiency and effectiveness of the team.
- Mentoring and training junior analysts to enhance their skills and knowledge.
- Collaborating with other teams, such as incident response and research, to ensure effective use of intelligence data in incident response and risk mitigation.

The work of an SOFC analyst, senior analyst, cybersecurity intelligence analyst, and senior cybersecurity intelligence analyst is fast-paced and dynamic. These roles focus on the collection, analysis, and dissemination of intelligence data to support effective cybersecurity operations and risk mitigation – which requires a combination of technical expertise, critical thinking skills, and effective communication and collaboration with other teams and stakeholders.

The integration of the SOFC's activities from the eyes on glass model to final remediation is critical for the success of the team. This process involves the continuous monitoring and detection of potential threats, followed by triaging and initiating response activities. The goal is to identify and contain any potential threats to the organization's network and data quickly and effectively. Monitoring and detection tools and methods are an important aspect of the SOFC's daily operations.

These tools provide the team with the ability to detect and monitor potential threat activity across the network. The art and science of effective monitoring and detection is a critical skill for SOFC analysts and team leaders. It requires a combination of technical expertise and critical thinking to be able to identify potential threats and take appropriate action. The SOFC makes effective use of monitoring and detection tools. To maximize the efficiency of these tools, the SOFC must:

- Continuously collect and analyze substantial amounts of data from various sources (network traffic, endpoints, cloud services, etc.).
- Use advanced analytics and machine learning algorithms to identify threats, anomalies, and suspicious activities.
- Implement real-time alerting and incident response procedures to quickly respond to potential threats and minimize damage.
- Continuously evaluate and improve monitoring and detection processes and tools.
- Maintain a highly skilled and knowledgeable team of cybersecurity experts to operate and maintain the tools and processes.

By utilizing a combination of technology, processes, and people, the SOFC can effectively detect and respond to cyber threats in a timely and efficient manner. This helps reduce the risk of a successful cyberattack and ensures the confidentiality, integrity, and availability of critical assets and data.

SOFC operations are both an art and a science because they require a combination of technical expertise and critical thinking skills to be effective. From a scientific perspective, cybersecurity SOFC operations are rooted in the application of technical and analytical tools such as network traffic analysis, endpoint protection software, and machine learning algorithms. These tools are used to identify potential threats and detect incidents. They rely on well-defined processes and procedures to ensure data are collected, analyzed, and acted upon in an efficient and effective manner. From an artistic perspective, cybersecurity SOFC operations require an elevated level of creativity, critical thinking, and problem-solving skills. Incidents often do not fit well-defined patterns; because of this, successful incident response requires creative thinking to identify and implement the best course of action. The constantly evolving threat landscape requires the ability to quickly adapt and adjust to new threats and techniques used by attackers. The combination of technical expertise and critical thinking skills is what makes cybersecurity SOFC operations both an art and a science. By effectively combining these skills and abilities, an SOFC can detect and respond to cyber threats in a timely and effective manner. This helps reduce the risk of a successful cyberattack and ensures the confidentiality, integrity, and availability of critical assets and data.

The cybersecurity analyst toolkit is a collection of tools and resources that SOFC analysts and team leaders use daily. These tools include various software applications, scripts, and command-line utilities that are used to monitor and detect potential threats. They also include various reference materials such as threat intelligence feeds, intrusion detection system, intrusion prevention system rules, and incident response plans. To maximize the efficiency and effectiveness of their work, a cybersecurity analyst toolkit should do the following:

- Include a comprehensive suite of cybersecurity tools. These tools should be able to collect, analyze, and visualize data from various sources, such as network traffic, endpoints, and cloud services. Some examples of tools that could be included in such a toolkit include:
- Network traffic analyzers such as Wireshark, netcat, or tcpdump.
- Endpoint protection solutions such as Windows Defender or McAfee.
- Cloud security tools such as AWS GuardDuty, Orca, or Google Cloud Security Command Center.
- Security Information and Event Management solutions such as Splunk or LogRhythm or Sentinel.
- Threat intelligence platforms such as ThreatConnect or Recorded Future.
- Comprehensive platforms such as Palo Alto or Tanium.
- Employ advanced analytics, such as machine learning algorithms, and artificial intelligence capabilities to detect anomalies, identify potential threats, and alert incident response teams.

- Implement effective incident response processes and procedures to ensure incidents are quickly and effectively contained and mitigated.
- Enable real-time collaboration and communication between teams to increase situational awareness and improve response times.
- Provide training and education to maintain an elevated level of knowledge and skills among team members.
- Continuously evolve to stay ahead of the rapidly changing threat landscape and to incorporate new tools and technologies as they become available.

By utilizing an effective cybersecurity toolkit, SOFC members can work together to detect, respond to, and mitigate cyber threats, reducing the risk of a successful cyberattack and ensuring the confidentiality, integrity, and availability of critical assets and data quickly and effectively.

This chapter covered the daily operations of the SOFC and the duties and responsibilities of the analysts, team leaders, and other personnel. It highlighted the importance of effective integration between monitoring and detection, triage, and remediation activities, and explained the role that the analyst toolkit plays in supporting these activities. The chapter began by discussing the integration of monitoring and detection activities with response activities – which includes the transition from the eyes on glass model to final remediation. It then explored the art and science of monitoring and detection tools and methods, including their effectiveness and limitations. The chapter concludes by discussing the analyst toolkit and the core elements that SOFC analysts and team leaders should have at their disposal. These core elements include the knowledge and skills necessary to effectively use monitoring and detection tools, as well as the critical thinking skills required to triage and respond to incidents.

6

FOUNDATIONS OF SECURITY
OPERATIONS OR SECOPS

SecOps, the guardian of our cyber realm, embraces the never-ending adventures of managing, monitoring, and preserving the safety of an organization's kingdom, treasures, and subjects.

Dr. Kevin Lynn McLaughlin, PhD

In this chapter we'll explore the complex nature of security operations, also known as SecOps, within the Cybersecurity Operations and Fusion Center (SOFC). SecOps refers to the ongoing processes and procedures used to manage, monitor, and maintain the security of an organization's infrastructure, data, and assets. This includes incident response, threat detection and analysis, security monitoring and reporting, and the deployment and maintenance of security technologies. SecOps is a critical component of the overall security posture of an organization. It encompasses the end-to-end remediation of security incidents, from alert, triage, and incident report to ticket handling and tracking. The SOFC is responsible for detecting and responding to security incidents/SecOps processes and procedures help to ensure these incidents are managed efficiently and effectively. This may include the use of automated tools and processes to detect and alert the team of cybersecurity incidents. SecOps also plays a critical role for a cybersecurity threat intelligence center (TIC). The TIC is responsible for collecting, analyzing, and disseminating threat intelligence information. For this book we are folding the TIC into the SOC and creating the SOFC. SecOps processes and procedures help ensure this information is used effectively to protect the organization. This may include the implementation of security controls to prevent the spread of malicious content, as well as the use of automated tools and processes to detect and respond to security incidents in real-time. Overall, SecOps is a critical component of the overall security posture

DOI: 10.1201/9781003259152-8   **45** for the cybersecurity SOFC. By employing well-defined processes and procedures, organizations can ensure that their SecOps are managed effectively and are able to respond quickly and efficiently to security incidents.

One of the key focus areas for SecOps is cybersecurity vulnerability management. This involves identifying, evaluating, and prioritizing vulnerabilities in an organization's information systems and networks, and then taking the appropriate measures to lessen or remediate those vulnerabilities. This process involves a continuous cycle of assessment, remediation, and monitoring to ensure that vulnerabilities are identified and addressed in a timely manner. Effective vulnerability management requires a thorough understanding of the organization's information systems and networks, as well as the software and hardware that support them. This includes identifying all potential points of entry for cyberattackers. These points of entry include software vulnerabilities, misconfigurations, or outdated software. Vulnerability management also involves understanding the potential impact of a successful attack on the organization's data, systems, and networks. Once vulnerabilities have been identified, they must be prioritized based on their potential impact, the likelihood of an attack, and how easy the vulnerability can be exploited by a threat actor. This allows organizations to focus their efforts and resources on the most critical vulnerabilities first.

Remediation strategies may include patching software, configuring systems to reduce the attack surface, or implementing additional security controls to minimize the risk. It is important to have a process in place to track the progress of remediation efforts, and to ensure that vulnerabilities are effectively eliminated or mitigated. Finally, ongoing monitoring is essential to ensure vulnerabilities do not reemerge, and new vulnerabilities are identified and addressed in a timely manner. This may include regular vulnerability scanning, penetration testing, and incident response activities.

Let's look at an example of running a cybersecurity vulnerability management program. It's possible to feel overwhelmed when you first arrive at an organization that has not invested in a major cybersecurity program, and you realize the amount of computer and network vulnerabilities in the environment. In many cases, it can be up to one million vulnerabilities that need to be remediated. In these instances, my advice is to keep a couple of concepts in mind: *kaizen*, or continuous improvement over time, and the old saying that you eat an elephant one bite at a time. As a practitioner who has faced this many times over the past 30 years, I can tell you that the task at hand is not as daunting to complete as it first appears.

The National Vulnerability Database (NVD) is part of the US government repository of standard-based vulnerability data and the National Institute of Standards and Technology. This vulnerability database enables automation of vulnerability management, enables security measurement, and ensures standard compliance. NVD contains a list of vulnerabilities, exposures, and an associated risk score for each vulnerability. This risk score is known as a common vulnerability and exposure (CVE) score and is an integral part of the NVD and Common Vulnerability Scoring System. Most vulnerability scanning tools and services make use of this CVE scoring system to assign a risk number or indicator to find vulnerabilities. The higher the CVE score, the greater the risk – meaning the vulnerability should be remediated more urgently.

I recall taking on an operational IT security role for one company. During my first morning, one of the cybersecurity team members came by with a cart. In the cart was a considerable number of papers – about 10,000 pages to be exact. The guy looked at me, said good morning, and then started to leave without taking the cart with him. I asked him why he was leaving the cart and he said "oh, those are all the vulnerabilities you need to tell IT to fix." I took a quick look and there were about 30 vulnerabilities listed per page. I took the cart to the loading dock, found a dumpster, and dumped all of the pages. Then I scheduled a meeting with the cybersecurity vulnerability team to talk about establishing a more effective process for handling vulnerabilities. The process we aligned on has proven effective across multiple companies. And it looks something like this: We started by agreeing that the risks listed in our current tool as catastrophic would be the most important ones to fix first. And that's because they would reduce the highest level of risk in the shortest amount of time. As a bonus, many of the patches that resolved the catastrophic risks would overlap in resolving some of the lower-level risks as well. At first there was always a staggering number of catastrophic risks to resolve, though not as staggering as a 10,000-page hardcopy PDF manual. We then agreed that we would deploy patches in the order that hit the greatest number of systems as opposed to deploying patches that only impacted one or two systems. In many cases a patch to resolve a vulnerability is needed across all Windows servers or all corporate firewalls instead of just one older version of the operating system. Once the catastrophic vulnerabilities were eliminated, we agreed that we would move on to the critical vulnerabilities, then the high vulnerabilities, then medium, and eventually we'd work our way down to the lower level vulnerabilities. We further aligned that the electronic report received by the vulnerability remediation team would show the following: system name, IP address, MAC address, OS level, vulnerability, and the recommended patch along with a status and comment column. Lastly, we aligned the number of patches that IT was willing to deploy during their weekly or monthly maintenance window. Following this approach, I was able to help reduce the number of catastrophic vulnerabilities across the infrastructure to nearly zero within 12 months. Some of you may question if that timeframe took too long, but keep in mind that some of these vulnerabilities had been left unattended and unpatched for years. The sheer volume of patching that IT was being asked to do was just overwhelming. At this point some readers might be questioning why IT would not just patch the systems regularly and why systems would be left unpatched with vulnerabilities. If you're wondering about this, you are correct! Automatic patching should be highly encouraged and should be used across most organizational systems. But the simple truth is that system patching can and does break an IT system, maybe not as much in modern days as in the past, but I can recall reading recently where OS and application patches from a trusted, and very large, vendor needed to be pulled back because they caused a major system failure. IT's job is to keep their organization up and running. Therefore, they are hesitant to complete broad-based patching for every issue on a weekly basis. By aligning on a risk-based approach that provides a manageable and workable solution for the IT patching team/support staff, we are more likely to get the cooperation needed to drive vulnerabilities downwards and reduce overall risk. IT leaders in the organization are very security conscious and they really do want to be great security stewards. They're also driven to keep the business

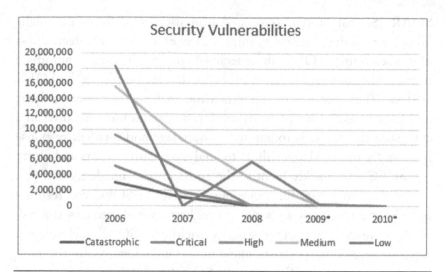

Figure 6.1 Vulnerability status.

Note: * Denotes steady state showing new vulnerabilities that are discovered and then removed at next patching cycle.

systems up and running and making money for the organization. As Stephen Covey said in his book, *The Seven Habits of Highly Effective People*, if we want to reduce risk, we need to work towards the win-win solution and then aggressively drive that forward while monitoring and adjusting the results as needed. By driving this one-team, one-goal approach to reducing risk through reduction of cybersecurity vulnerabilities (Figure 6.1), the chances of a successful outcome are maximized. Do this work right and you end up with a metric that looks like the one above, which is a remarkable story to share with your executive leadership team.

In summary, cybersecurity vulnerability management is a critical aspect of protecting the organization's assets, systems, and networks. It involves identifying, assessing, and mitigating vulnerabilities that could be exploited by cyberattackers. It also involves monitoring the progress of remediation efforts and ensuring that vulnerabilities are effectively eliminated or mitigated.

Automation capabilities play a vital role in SecOps, as they enable the SOFC team to manage and respond to security incidents efficiently and effectively. Automation, artificial intelligence (AI), and

SOAR (Security Orchestration, Automation, and Response) technologies have the potential to improve the efficiency and effectiveness of a cybersecurity SOFC. These technologies can be used to automate many of the routine and repetitive tasks that are performed by SOFC analysts. They free up the analysts' time to focus on more complex and critical tasks. One of the ways that automation can be used in a cybersecurity SOFC is to automate the triage and incident response process. By using AI algorithms to analyze security events and alerts, the SOFC can quickly identify the most critical incidents and prioritize their response. This helps ensure that the SOFC can respond quickly and efficiently to security incidents, which reduces the risk of data breaches and other security incidents. SOAR technologies can also be used to automate many of the tasks involved in the incident response process. For example, SOAR technologies can be used to automate the data gathering process of a security incident, analyze that data to determine the cause of the incident, and then deploy the appropriate response. This helps ensure that the incident response process is efficient and effective, and that the SOFC can respond quickly and effectively to cybersecurity incidents. Cybersecurity automation refers to the use of technology, such as software and algorithms, to automate and streamline various cybersecurity processes and tasks. This can include tasks such as vulnerability management, incident response, and threat detection and response. The goal of cybersecurity automation is to increase efficiency, reduce human error, and enhance the overall security posture of an organization. Another example of cybersecurity automation is the use of automation tools for patch management. These tools can automatically identify vulnerabilities, assess the risk to the organization, and deploy patches to remediate those vulnerabilities. This not only saves time for security teams, but also helps ensure that vulnerabilities are addressed in a timely manner, reducing the risk of a successful attack. Furthermore, cybersecurity automation should effectively make use of AI and machine learning to detect and respond to cyber threats. These technologies can analyze substantial amounts of data and identify patterns that may indicate a potential threat. They can also automatically take actions to remediate the threat. These actions include blocking network access or quarantining a compromised device. It is important to note that while cybersecurity automation can bring many benefits, it should be used in conjunction with human oversight and expertise. Automation tools should be configured and monitored by security professionals who can assess the results of the automation and make decisions based on their knowledge and expertise. Cybersecurity automation is an important aspect of modern SecOps. It can help organizations improve their security posture by automating repetitive and tedious tasks – which allows security teams to focus on more high-priority and complex tasks, such as threat hunting and incident response.

As mentioned earlier in this chapter, the use of cybersecurity SOAR, tools, and processes is a great complimentary technology for automation that helps create a cohesive system that automates repetitive and time-consuming tasks. This allows security teams to focus on more high-level, strategic tasks, such as threat hunting and incident response. SOAR platforms typically include a combination of orchestration, automation, and incident response capabilities. Orchestration is the process of coordinating the actions of multiple security tools and systems to automate workflows and streamline incident response. Automation is the use of software and algorithms to perform repetitive tasks and process large volumes of data. Incident response refers to the process of identifying, containing, and mitigating security incidents. Effective use of SOAR platforms can automate a wide range of security tasks, including threat intelligence gathering, vulnerability management, incident triage, and incident response. This allows security teams to be more efficient, effective, and proactive in their efforts to protect their organization's assets. Additionally, SOAR platforms can provide detailed reporting and analytics capabilities. This allows security teams to track and measure the effectiveness of their SecOps and make data-driven decisions to improve their overall security posture.

Now let's take a look at the SOFC cybersecurity threat intelligence in reference to how automation, AI, and SOAR technologies and processes can be used in the TIC process to increase overall mission effectiveness. Automation, AI, and SOAR technologies can play a critical role in enhancing the ability of an SOFC to gather, analyze, and disseminate threat intelligence. These technologies can help the SOFC focus on more complex and critical tasks by automating many of the routine and repetitive tasks involved in the threat intelligence process. Some of these more complex tasks include identifying emerging threats and providing actionable intelligence to the wider organization. One of the ways that automation can be used in an SOFC is to automate the collection and analysis of threat intelligence data. For example, AI algorithms can be used to analyze substantial amounts of security data, identify patterns and correlations, and generate insights into emerging threats. This helps ensure the SOFC can quickly and efficiently identify threats, thereby reducing the risk of data breaches and other security incidents. SOAR technologies can also be used to automate the dissemination of threat intelligence. For example, SOAR technologies can be used to automate the process of distributing threat intelligence to relevant stakeholders within the organization. This helps ensure the SOFC threat intelligence function can distribute threat intelligence quickly and efficiently, which reduces the risk of data breaches and other security incidents. Automation, AI, and SOAR technologies and processes have the potential to increase the overall effectiveness of a cybersecurity SOFC by automating many of the routine and repetitive tasks involved in the threat intelligence process. These technologies can help the SOFC identify and disseminate threat intelligence, reducing the risk of data breaches and other security incidents quickly and efficiently.

Proactive threat hunting is another key component of SecOps. The SOFC threat hunting team must be able to proactively identify and counteract potential threats before they can cause damage to the organization. Cybersecurity threat hunting is the proactive and dynamic process of searching for, discovering, and mitigating cyber threats that may have dodged the traditional security controls. Threat hunting teams, which consist of experienced cybersecurity analysts, utilize a combination of techniques like data analysis, threat intelligence, and manual investigation to uncover potential threats and vulnerabilities in the organization's network. The goal of threat hunting is to identify and isolate malicious activity before it causes significant harm to the organization. To achieve this, threat hunting teams search for signs of known malware and indicators of compromise. They also analyze network traffic for unusual patterns and review system logs for unusual activity. Threat hunting also involves the use of specialized tools and technologies like endpoint detection and response (EDR) systems to assist in the identification and investigation of potential threats. Threat hunting is a big part of a thorough security strategy and enhances the organization's ability to detect, respond to, and prevent cyberattacks. As mentioned previously, the threat hunting process uses a combination of data analysis, threat intelligence, and manual investigation techniques to uncover potential threats and vulnerabilities in the network. These techniques include searching for signs of known malware and indicators of compromise, analyzing network traffic for unusual patterns, reviewing system logs for unusual activity, and employing specialized tools and technologies. These special tools and technologies include EDR systems that assist in the identification and investigation of potential threats. Threat hunting teams work closely with other teams and departments to ensure a collaborative and effective response to potential threats. Threat hunting is a critical component of an organization's overall security strategy, providing a proactive approach to detecting and mitigating cyber threats. It focuses on discovering and isolating malicious activity. Threat hunting teams play a crucial role in improving an organization's ability to prevent and respond to cyberattacks.

In conjunction with the threat hunting team, the SOFC team may also want to consider employing the Red Team and Purple Team. The Red and Purple Teams are useful because they simulate real-world attacks and test the organization's defenses. A Red Team is a simulated adversarial group that tests an organization's security defenses by simulating real-world attacks. The Red Team's goal is to identify weaknesses and vulnerabilities in the organization's security posture. Once the weaknesses and vulnerabilities are identified, the Red Team helps the organization better understand how to prevent, detect, and respond to these types of attacks. Red Team operations typically involve a range of techniques like social engineering, physical security testing, and technical exploitation to identify and exploit any security gaps. A Purple Team is a collaborative effort between the organization's security defenders (Blue Team) and the Red Team. The two teams work together to identify and remediate security weaknesses. The Purple Team approach allows the organization to benefit from the expertise and skills of both its SecOps team and its simulated attackers. The goal of the Purple Team is to improve the overall security posture of the organization by incorporating insights and providing training on lessons learned from Red Team operations to the Blue Team analyst. These teams are an essential part of creating strong cybersecurity strategies and processes for an organization. In terms of process and methods, both Red Teams and Purple Teams make use of a variety of tools and techniques. These techniques include data analysis, threat intelligence, and specialized security tools, such as EDR systems. Additionally, both teams benefit from effective collaboration and communication. They also benefit from a thorough understanding of the organization's cybersecurity posture and the tactics, techniques, and procedures used by real-world attackers.

Cyber intelligence gathering is the process of collecting, analyzing, and distributing information that pertains to potential cyber threats and vulnerabilities. This can include information about known and unknown threat actors, their tactics, techniques, and procedures, as well as information about the tools and infrastructure they use to conduct cyberattacks. The goal of cyber intelligence gathering is to provide organizations with the information they need to improve their overall security posture and to better protect themselves against cyber threats. Cyber intelligence gathering can take many forms. These forms include the collection of open-source information, the use of proprietary intelligence feeds and cyber reconnaissance operations. The collection of open-source information can include monitoring social media, online forums, and other publicly available sources for information about cyber threats and vulnerabilities. Proprietary intelligence feeds can include data from vendors, partners, and other organizations that specialize in the collection and analysis of cyber threat intelligence. Cyber reconnaissance operations can include the use of tools and techniques to gather information about a target's network infrastructure, systems, and users. The information is then analyzed to determine its relevance and credibility. This process can involve the use of link analysis, network analysis, and data visualization to identify patterns and connections between different pieces of information. The information is then distributed to various groups located in the SOFC such as the incident response teams, threat hunting teams, operational leaders, and executives to aid in the detection, response, and lessening of cyber threats. Cyber intelligence gathering is a critical component of an organization's overall security strategy. It provides the organization with the information it needs to identify and respond to potential cyber threats in a timely and effective manner. It is also important to note that the cyber intelligence gathering is not just a one-time event, but rather a continuous process that is updated regularly with current and emerging cyber threats.

Risk and threat reporting is a critical aspect of SecOps. The SOFC team must be able to provide accurate and timely information on the organization's risk posture and any potential threats to the organization's leadership. Tool support is an integral part of SecOps. The SOFC must be able to efficiently and effectively manage the various security tools deployed by the organization. This may include adding dedicated IT SecOps team members with specific expertise in managing and maintaining these tools. Cyber threat and risk reporting refers to the process of identifying, analyzing, and communicating information about potential cyber threats and vulnerabilities that may impact an organization. This includes gathering data from network logs, intrusion detection systems, and threat intelligence feeds, and analyzing that data to identify potential threats and vulnerabilities. The resulting information is then used to inform decision-making within the organization. Those decision-making efforts can include prioritizing security efforts and allocating resources to address the most critical threats. The process of cyber threat and risk reporting is a critical component of any organization's security program. It enables organizations to stay informed about the latest threat trends and tactics, as well as identify and track specific threats that may be targeting their networks. This knowledge is essential for effective incident response and threat management, and for making informed decisions about security investments and priorities. Effective cyber threat and risk reporting requires a combination of technical expertise, analytical skills, and an understanding of the larger threat environment. Additionally, it requires the ability to effectively communicate complex technical information to non-technical stakeholders, such as senior management and business leaders. To sum it up, cyber threat and risk reporting is a vital process that enables organizations to stay informed about the latest cyber threats and vulnerabilities. It also helps organizations make informed decisions about how to best protect their networks and data. It requires a combination of technical expertise, analytical skills, and an understanding of the broader threat landscape and the ability to communicate the results of the analysis to a non-technical audience.

This chapter focused on SecOps and its essential role in the SOFC. We explored the end-to-end remediation process, including alert triage, security incident reporting, ticket handling, and tracking. It also highlighted the importance of vulnerability management, automation capabilities, cyber intelligence gathering, threat hunting, and the role of the Red and Purple Teams in securing the organization. Additionally, the chapter covered the importance of risk/threat reporting and tool support, including the potential inclusion of cross functional SecOps team members in the cybersecurity SecOps process. Hopefully after reading this chapter, you will be able to take away a comprehensive overview of the critical components of SecOps and its role in elevating the SOFC's ability to effectively detect, respond to, and prevent cyberattacks.

7

DETECTION, RESPONSE, AND REMEDIATION

Being a cybersecurity defender is a task that requires bravery and fortitude; it takes brains and courage to excel as a cyber defender.

Dr. Kevin Lynn McLaughlin, PhD

In this chapter we will focus on the core tasks and responsibilities of cybersecurity analyst, cybersecurity senior analyst, cyber threat intelligence analyst, and cyber threat intelligence senior analyst within the Cybersecurity Operations and Fusion Center (SOFC). Some of the topics we'll touch on in this chapter include cybersecurity incident response, incident and breach containment, remediation, forensic analysis, return to normalcy, and resiliency as it pertains to cybersecurity. We will also discuss the talent, practice events, and live operations associated with the cyber incident response team (CIRT). Having an effective response to what has been detected is a core task and critical skill for SOFC members. This chapter discusses the details of effective response for incidents of all sizes. It covers how some items can and should be automated, delegated, etc., and how some need a full CIRT response. The chapter goes on to cover how a CIRT is staffed and how it operates.

Cybersecurity Incident Response

Effective cybersecurity incident response for an SOFC refers to a systematic and coordinated approach for managing and resolving security incidents in a timely manner. This process involves identifying and assessing the threat, containing and eradicating the issue, and restoring normal operations. An SOFC performs incident response by following a structured and methodical approach that is designed to

DOI: 10.1201/9781003259152-9

address security incidents in a timely and effective manner. The incident response process typically involves the following steps:

- **Preparation:** Establish incident response policies, procedures, and protocols, and conduct regular training and exercises to ensure readiness for real-world incidents.
- **Detection:** Monitor various security systems and sources for signs of security incidents, and promptly raise alerts to initiate the response process.
- **Analysis:** Quickly assess the nature, scope, and severity of the incident, and gather relevant information to inform decision-making.
- **Containment:** Implement measures to isolate the affected systems and prevent the spread of the incident.
- **Eradication:** Identify and remove the root cause of the incident to restore normal operations.
- **Recovery:** Bring the affected systems back online and conduct a post-incident review to identify areas for improvement.
- **Lessons Learned:** Capture and analyze lessons learned from the incident to help implement future incident response planning and strategies.

The SOFC and other relevant security teams can assist each other quickly and efficiently. As an example, the CIRT can be supported by a threat intelligence center (TIC) analyst through the provisioning of real-time intelligence and insights about the threat actors, their tactics, techniques, and procedures. This information helps the CIRT quickly identify the scope and nature of the incident, prioritize the response actions, and implement the appropriate countermeasures to prevent future incidents. The SOFC's ability to provide a centralized repository of security incidents and analysis is essential. That's because the repository can be used to detect patterns and trends, and help the team create future incident response plans and strategies. The SOFC plays a crucial role in defending against cyber threats by performing incident response, which requires a combination of technical expertise and situational awareness. The SOFC security teams focus on monitoring, detection, response, threat intelligence, etc. These teams work collaboratively to provide the organization with a well-defined and tested process. A well-functioning intelligence function helps the incident responders enhance the effectiveness and efficiency of their cybersecurity incident response.

Breach Handling and Containment

Effective breach and containment in cybersecurity involves quickly identifying and isolating a security incident to prevent further damage and spread. The following steps are key in achieving effective breach and containment:

- **Identification:** Detect the breach using monitoring systems and incident response protocols.
- **Containment:** Isolate the affected systems and networks to prevent the spread of the breach.
- **Analysis:** Investigate the cause and extent of the breach to determine the best course of action.
- **Remediation:** Implement measures to prevent future breaches and repair any damage done during the incident.

The SOFC plays a crucial role in supporting this process by providing timely and accurate information on the latest threats. This helps organizations respond quickly and effectively to security incidents. The SOFC also provides a centralized repository of threat data that can be used to proactively defend against potential breaches. Effective breach and containment in cybersecurity requires a multistep process that includes identification, containment, analysis, and remediation. The SOFC is an essential component of this process because it provides critical threat intelligence and information to support quick and effective responses to security incidents.

Remediation

Effective cybersecurity incident remediation involves repairing any damage done during a security breach and implementing measures to prevent future incidents. The following steps are key in achieving effective remediation:

- **Root Cause Analysis:** Determine the underlying cause of the breach to prevent future incidents.

- **Damage Assessment:** Evaluate the extent of the damage caused by the breach.
- **Remediation:** Implement measures to repair the damage and prevent future incidents. These measures include patching vulnerabilities, enhancing security systems and procedures, and updating employee training programs.

It is important to understand that the response to a cybersecurity incident does not follow the standard IT practice of restoring service immediately. While restoring systems as quickly as possible is desirable, this should not be done at the cost of allowing a cyber incident to escalate or a threat actor to stay active within an organization's environment. Unfortunately, organizations without mature cybersecurity programs in place do not recognize this. These organizations may experience an unusual event and allow their IT personnel to turn off the system's anti-malware controls – as they think doing so will solve the issue. However, disabling critical cybersecurity control that has identified security issues during an ongoing attack is not an effective way to protect the organization's data or assets. I have seen far too many cases where ransomware and other forms of malware were allowed to spread throughout the organization's infrastructure because the IT team responded by turning off crucial security controls. This is why it's important to have a CIRT in place to handle cybersecurity incidents, as opposed to allowing the IT team to handle these occurrences. The SOFC plays a crucial role in supporting this process by providing real-time threat intelligence and information on the latest threats, vulnerabilities, and attack techniques. This information enables organizations to quickly identify and address the root cause of the breach, which is essential for preventing future incidents. The SOFC also provides a centralized repository of threat data that can be used to continuously improve an organization's security posture. Effective remediation in cybersecurity requires a comprehensive approach that includes root cause analysis, damage assessment, and remediation. The SOFC is an invaluable resource for the remediation process because the SOFC provides critical threat intelligence and information to enable organizations to respond to security incidents and prevent future breaches quickly and effectively.

Forensic Analysis as Part of Cybersecurity Incident Response

Effective cybersecurity incident forensic analysis provides valuable insights into the cause, extent, and impact of a security breach. The following are the benefits of conducting effective forensic analysis:

- **Identification of Root Cause:** Understanding the underlying cause of the breach enables organizations to prevent future incidents.
- **Evidence Preservation:** Collecting and preserving evidence helps organizations to effectively respond to the incident and support legal actions.
- **Improved Security Posture:** Gaining a comprehensive understanding of the breach enables organizations to continuously improve their security posture.

The process of conducting forensic analysis involves collecting and preserving evidence. From there, it's important to analyze the data to determine the cause and extent of the breach and use the findings to improve the organization's security posture. The forensic analysis process typically includes the following steps:

- **Collection:** Gathering and preserving evidence from affected systems and networks.
- **Analysis:** Examining the data to determine the cause and extent of the breach.
- **Reporting:** Presenting the findings in a clear and concise manner to support decision making and continuous improvement.

A cybersecurity SOFC plays a key role in supporting the forensic analysis process by providing real-time threat intelligence and information on the latest attack techniques and vulnerabilities. This information can be used to guide the forensic analysis and identify the root cause of the breach. The SOFC also provides a centralized repository of threat data that can be used to continuously improve an organization's security posture. Effective cybersecurity incident forensic analysis provides valuable insights into the cause and impact of a security breach, supports evidence preservation and legal actions, and enables organizations to continuously improve their security posture. The SOFC is a key component in supporting the forensic analysis process by providing timely and accurate threat intelligence and information.

Return to Normal Operations

Return to normal operations (RTO) is the process of restoring normal business operations following a security breach. The following are the benefits of conducting effective RTO:

- **Minimizing Downtime:** Restoring normal operations as quickly as possible minimizes the impact of the breach on business operations.
- **Confidence Building:** Quickly returning to normal operations builds confidence in the organization's ability to respond to security incidents.
- **Reputation Protection:** Effective RTO helps minimize the impact of the breach on the organization's reputation.

The process of conducting RTO typically involves the following steps:

- **Containment:** Isolating the affected systems and networks to prevent the spread of the breach.
- **Remediation:** Implementing remediation activities to eliminate the root cause of the breach.
- **Validation:** Testing and validating the remediation activities to ensure that the systems and networks are secure.
- **Recovery:** Restoring normal business operations.

A cybersecurity TIC within an SOFC plays a key role in supporting the RTO process by providing real-time threat intelligence and information on the latest attack techniques and vulnerabilities. This information can be used to guide containment and remediation activities, and to validate their effectiveness. The TIC also provides a centralized repository of threat data that can be used to continuously improve an organization's security posture.

It's important for SOFC members to understand what path a hacker used to exploit an organization's vulnerabilities when they begin to return systems to normal operations. Understanding the pathway that a hacker takes enables the cybersecurity team to prevent similar incidents from occurring in the future. It also helps organizations improve their cybersecurity posture by identifying any vulnerabilities that may have been exploited by the attacker and implementing remediation activities to eliminate those vulnerabilities. Effective cybersecurity incident RTO minimizes downtime, builds confidence, and protects an organization's reputation. A cybersecurity TIC plays a critical role in supporting the RTO process by providing real-time threat intelligence and information, and by enabling organizations to continuously improve their security posture. Understanding the path a threat actor takes is a key aspect of returning to normal operations as it helps organizations to prevent similar incidents from occurring in the future.

Resiliency

SOFC team members need to recognize and understand that cyber resiliency is a critical concept that must be matured within their organization. Cyber resiliency is the ability of an organization to quickly recover from a cyberattack and continue its normal business operations. The following are the key components of cyber resiliency:

- **Preparedness:** Organizations must have plans, procedures, and processes in place to respond to a cyberattack.
- **Detection:** Organizations must be able to detect cyberattacks in real-time so that they can respond quickly.
- **Response:** Organizations must be able to respond quickly and effectively to cyberattacks.
- **Recovery:** Organizations must be able to quickly recover from cyberattacks and return to normal business operations.

Achieving cyber resiliency involves the following steps:

- **Assessment:** Organizations must assess their current security posture and identify any gaps.
- **Planning:** Organizations must develop plans, procedures, and processes to respond to cyberattacks.

- **Implementation:** Organizations must implement the plans, procedures, and processes that have been developed.
- **Testing:** Organizations must evaluate their preparedness, detection, response, and recovery capabilities to ensure that they are effective.

Cyber resiliency is the ability of an organization to quickly recover from a cyberattack and continue its normal business operations. The process of achieving cyber resiliency involves assessment, planning, implementation, and testing. The SOFC plays a critical role in supporting the process by providing real-time threat intelligence and information and enabling organizations to continuously improve their security posture.

Cyber Incident Response Team

When an organization experiences a major cyber incident, they need to have a well-trained cross-functional CIRT to call on. This team should be composed of highly trained and passionate personnel who come from all aspects of the business, including IT departments and supporting functions such as Legal, HR, Finance, Communications, Product Cybersecurity, Corporate Security, and any other teams that are applicable to the organization. The CIRT is a group of cybersecurity professionals within an organization who are responsible for detecting, responding to, and mitigating cybersecurity incidents. CIRT members have the technical expertise and knowledge to investigate and resolve cybersecurity incidents quickly and effectively. The CIRT is typically staffed by a combination of employees from the IT, security, and legal departments. This cross-functional team is essential to ensure that all aspects of an incident are addressed, from technical mitigation to legal and regulatory compliance. When a major cybersecurity incident occurs, the CIRT will activate its incident response plan. It is advisable to have a trusted cybersecurity partner on retainer who can provide additional expertise and resources as needed. If you do make use of a cybersecurity partner, they should be included in a few of the CIRT practices that take place throughout the year so that they have a solid understanding of their role in your CIRT and how your CIRT works when handling an incident. Activating the CIRT typically involves the following steps:

- **Containment:** The CIRT will take immediate steps to contain the incident and prevent it from spreading.
- **Investigation:** The CIRT will investigate the incident to determine its cause, scope, and impact.
- **Analysis:** The CIRT will analyze the data collected during the investigation to determine the source and nature of the attack.
- **Response:** The CIRT will take action to mitigate the incident. This includes removing malware, restoring systems, and updating security controls.
- **Recovery:** The CIRT will work to restore normal business operations and ensure that the organization's systems and data are secure.
- **Lessons Learned:** The CIRT will document lessons learned from the incident to improve the team's response and effectiveness the next time they are called into action.

The CIRT is responsible for detecting, responding to, and mitigating cybersecurity incidents. To effectively fulfill this role, CIRT members are assigned specific responsibilities. The following is a list of common roles and responsibilities found within a CIRT:

- **Incident Commander:** This individual is responsible for overseeing the entire incident response process and making key decisions.
- **Technical Lead:** This individual is responsible for leading the technical aspects of the response, such as conducting investigations, analyzing data, and removing malware.
- **Communication Lead:** This individual is responsible for managing communication during the incident response, including providing updates to stakeholders, communicating with law enforcement, and managing the release of information.
- **Documentation Lead:** This individual is responsible for documenting the incident response process and maintaining an accurate record of all actions taken during the response. Note – I usually call the resources on point for documenting the event Scribes.

- **Containment Lead:** This individual is responsible for ensuring that the incident is contained and helping to prevent it from spreading.
- **Recovery Lead:** This individual is responsible for restoring normal business operations and ensuring that the organization's systems and data are secure.
- **Forensics Lead:** This individual is responsible for conducting a forensic analysis of the incident and determining the cause, scope, and impact of the attack.
- **Threat Intelligence Lead:** This individual is responsible for providing real-time threat intelligence and information on the latest attack techniques and vulnerabilities.

These are common roles and responsibilities found in a CIRT. However, the specific roles and responsibilities tend to vary depending on the size and complexity of the organization, and the nature of the specific incident that must be responded to. The important thing is to ensure that the CIRT has the necessary resources and personnel to effectively respond to incidents. The SOFC plays a critical role in supporting the CIRT during the incident response process. The SOFC TIC provides real-time threat intelligence and information on the latest attack techniques and vulnerabilities. This information can be used to guide the CIRT's investigation and response activities – which enables them to quickly identify and address the root cause of the incident. The TIC also provides a centralized repository of threat data that can be used to support incident response activities. This repository helps ensure that the CIRT has the information and resources it needs to resolve incidents quickly and effectively.

Consistent training and practice are essential for establishing and maintaining an effective CIRT. It is critical that the team is prepared and ready to act within a moment's notice. Regular training and practice allow the CIRT to identify and address any weaknesses in their response processes. This helps improve their skills and knowledge and helps ensure they can effectively respond to incidents. CIRT members should practice their response processes and procedures on a regular basis – every quarter, ideally. This will help them stay current on the latest threat information and techniques, as well as ensure they're familiar with their roles and responsibilities in the event of an incident. Regular practice allows the CIRT to identify any potential obstacles or bottlenecks in their response processes and make necessary adjustments. Furthermore, it is important to have an independent reviewer assess the CIRT's practices and processes. This ensures that any gaps or areas for improvement are identified and addressed, and it provides a valuable third-party perspective on the effectiveness of the CIRT's response processes. An independent reviewer can also provide objective recommendations for improvement and can help ensure that the CIRT is prepared to respond to the latest threats and attack techniques. Regular CIRT training and assessment is critical for ensuring that the team is prepared and ready to respond to incidents. The CIRT should practice their response processes and procedures on a regular basis and have their practices evaluated and assessed by an independent reviewer to identify any weaknesses or areas for improvement. This helps ensure that the organization is protected from cyber threats and that the CIRT can effectively respond to incidents when they occur. Further, it is important that an organization regularly incorporates tabletop exercises as part of a cybersecurity CIRT training program. These exercises should involve the organization's executive team and provide an opportunity for them to experience simulated cybersecurity scenarios and respond to potential threats in a controlled environment. The executive tabletop exercises can help build the executive team's understanding of the current threat landscape, and their role in the response to a security incident. For example, tabletop exercises could simulate a phishing attack, a data breach, or a ransomware incident. The executive team can then practice responding to the simulated scenario and engage in discussion and decision-making processes related to the incident. This type of hands-on training can help build the team's confidence in their ability to respond to security incidents and increase their understanding of the importance of a comprehensive security program. By regularly incorporating these exercises into the training program, the cybersecurity CIRT can help create a strong security culture and ensure the organization is prepared to effectively respond to potential threats. Whenever and as much as possible the CIRT training events should be umpired and graded by an independent third party who provides improvement feedback to the CIRT leaders.

In summary, a CIRT is a cross-functional group of expert resources within an organization who are responsible for responding to and mitigating cybersecurity incidents. The CIRT operates by activating its incident response plan in the event of a major incident, which typically involves containment, investigation, analysis, response, and recovery. A cybersecurity SOFC plays a critical role in supporting the CIRT by providing real-time threat intelligence and information and helps enable the CIRT to resolve incidents quickly and effectively.

PART III
REPORTING AND METRICS

8
SOFC REPORTING

To be a successful cybersecurity leader, one must possess the ability to craft captivating narratives that ignite the imagination of executive decision-makers.

Dr. Kevin Lynn McLaughlin, PhD

Reporting

It is critical that the Cybersecurity Operations and Fusion Center (SOFC) tells its story to organizational leaders. To help tell that story, the SOFC should provide daily, weekly, monthly, and annual cybersecurity reports that drive continuous improvement towards the effectiveness of the organization's cybersecurity efforts. Here is a list of a few of the cybersecurity reports to be considered:

- **Security Operations** – These operational reports should include real-time monitoring and threat intelligence data, security incidents and alerts, and their resolution status. Executive level reports should highlight trends, threats, and risk exposure levels.
- **Red Team** – These operational reports should document simulated attack scenarios and their success rate, as well as any identified gaps or vulnerabilities. Executive level reports should provide a high-level overview of the effectiveness of the organization's security posture and any recommended improvements. As an item of interest to organizational leaders, any use of artificial intelligence (AI) or machine learning (ML) technologies to create or deploy Red Team events should be clearly called out.
- **Threat Hunting Team** – These operational reports should include details of proactive threat hunting activities like

DOI: 10.1201/9781003259152-11

identifying and tracking threats and documenting any malicious activities or threat actors. Executive level reports should summarize the overall threat landscape and the organization's ability to detect and respond to threats. As an item of interest to organizational leaders, any use of AI or ML technologies to create or deploy Red Team events should be clearly called out.

- **Vulnerability Management Team** – These operational reports should document the progress of vulnerability scans, the number and severity of vulnerabilities found, and the status of their resolution. Executive level reports should summarize the overall risk posture of the organization's systems and applications.

- **SOAR Automation Team** – These operational reports should detail the use of security orchestration, automation, and response (SOAR) technology to automate security processes and reduce incident response times. Executive level reports should highlight the effectiveness of the SOAR system in reducing incident response times and improving the overall security posture. If any forms of AI or ML technologies are being used to create cybersecurity automations these should be called out in the report; along with the number of hours leveraging these tools are saving the automation team members.

- **Cybersecurity SOFC** – These operational reports should highlight any areas for improvement and provide recommendations for future investments in cybersecurity. Executive level reports should provide a comprehensive view of the organization's overall cybersecurity posture, including current threats, vulnerabilities, and risk levels.

It is important for these teams to regularly communicate and share their reports. This allows for a more comprehensive and effective approach to securing the organization's assets. It's important that the reports produced by the SOFC are shared with cross-functional IT team leaders. The SOFC should be sharing operational reports that detail the current threat landscape, any security incidents and their resolution status, and real-time monitoring and threat intelligence data. This information will provide the IT team leaders with a clear understanding of the organization's current security posture and allow them to make informed decisions regarding their own systems and applications. While the SOFC should be producing reports that summarize the overall cybersecurity posture of the organization, these reports should also provide recommendations for future investments in cybersecurity and highlight any areas for improvement. This information will help IT team leaders prioritize their cybersecurity efforts and ensure that their systems and applications are secure and aligned with the organization's overall cybersecurity strategy. It is important for the SOFC to regularly communicate with cross-functional IT team leaders, as this will create more collaboration and ensure that all IT systems and applications are properly secured. By providing regular and clear reports, the SOFC can help IT team leaders stay informed and engaged, which will lead to a stronger and more secure organization. The SOFC should share joint IT infrastructure and cybersecurity metrics with the C-Suite and Board of Directors. Here are a few metrics that can included on this type of report:

- **Cybersecurity Incidents** – number and types of incidents, and time taken to resolve.
- **Cybersecurity Risk Reduction** – measures taken to reduce risk and their impact.
- **Compliance Status** – updates on adherence to relevant laws and regulations, such as GDPR, HIPAA, etc.
- **Network and System Availability** – uptime, downtime, and mean time to repair.
- **Budget** – spending on cybersecurity and IT infrastructure, as well as budget utilization.
- **Disaster Recovery** – readiness and performance of disaster recovery plans.
- **Personnel and Training** – number of staff, their qualifications and training, and turnover rates.
- **Technology Upgrades and Replacements** – plans and progress of technology upgrades and replacements.

It is important for cybersecurity leaders and IT infrastructure leaders to share agreed upon joint metrics with the C-Suite and Board of Directors. The purpose of sharing these metrics is to communicate the effectiveness of the organization's cybersecurity and IT infrastructure efforts, and to provide an overall view of the organization's security posture. It is also important to show senior executives that there is a strong partnership between the cybersecurity and IT infrastructure team regarding protection from cyberattacks. The C-Suite and Board of Directors can obtain a clear understanding of the organization's current state and make informed decisions regarding investments in cybersecurity and IT infrastructure when they have a set of impactful metrics in front of them. These metrics serve as a baseline for measuring progress and highlighting areas for improvement. They help align the organization's cybersecurity and IT infrastructure efforts with the organization's overall business and cybersecurity strategy and ensure that the organization's resources are being used effectively to reduce risks and protect critical assets. By sharing these metrics with the C-Suite and Board of Directors, cybersecurity and IT infrastructure leaders are demonstrating their commitment to transparency, accountability, and good governance. The use of proper metrics will help build trust and confidence in the organization's ability to manage risks and secure its assets. The sharing of agreed upon joint metrics between cybersecurity and IT infrastructure leaders with the C-Suite and Board of Directors is critical for effective risk management, alignment with business strategy, and accountability.

Reporting Frequency

The reporting frequency for the reports produced by the SOFC should be determined based on the organization's specific needs and requirements. Here are some general guidelines that can be followed:

- **Executive Management:** The frequency of reports for executive management should be driven by the level of detail required and the criticality of the information. Typically, executive management may require high-level, summarized reports on a monthly or quarterly basis, with additional reports being provided as necessary in response to major incidents or critical events.
- **Operational Reports:** The frequency of operational reports should be determined based on the operational needs of the organization. For example, real-time threat intelligence

and incident response reports may need to be generated on a daily or weekly basis. Monthly reports detailing the overall security posture may be more appropriate. The goal is to provide enough information to support continuous improvement of the cybersecurity program, while not overwhelming the organization with excessive information. In general, the reporting frequency should be tailored to meet the specific needs of the organization and should be designed to support continuous improvement and the effectiveness of the cybersecurity program.

- **Annual Reports**: Annual reports should be used to show how effective the Cybersecurity SOFC has been operating over an extended period. Here are the reasons for sharing reports on an annual basis:
  - **Assessment of Progress**: Annual reports provide a comprehensive view of the organization's cybersecurity posture over a longer timeframe. This allows executive leadership to assess the progress made in improving the organization's security posture and make decisions about future investments in cybersecurity.
  - **Alignment with Business Strategy**: Annual reports help align the organization's cybersecurity efforts with the overall business strategy. By providing a clear picture of the organization's security posture and the measures being taken to reduce risks, executive leadership can make informed decisions about the allocation of resources and ensure that the organization's cybersecurity efforts are aligned with the overall business strategy.
  - **Compliance**: Many organizations are subject to regulatory requirements that mandate regular cybersecurity reporting. Annual reports provide a consolidated view of the organization's compliance with these requirements. This helps ensure the organization is meeting its legal obligations.
  - **Transparency and Accountability**: Providing annual reports to executive leadership demonstrates the organization's commitment to transparency and accountability. This helps build trust and confidence in the organization's ability to manage risk and secure its assets.

Annual Cybersecurity SOFC metrics and reporting play a critical role in assessing progress, aligning with business strategy, ensuring compliance, and promoting transparency and accountability. These reports are essential for providing executive leadership with the information they need to make informed decisions about the organization's cybersecurity posture and the measures being taken to reduce risks in the future.

Tailoring and Leveraging Reports to Drive Success

It is imperative for cybersecurity leaders to comprehend the impact that effective reporting can have on enhancing security within their organization. Reports serve to increase awareness of vulnerabilities and expedite resolution of discovered issues. However, keep in mind that the reports should not be excessive. Instead, they should be focused and targeted towards achieving specific outcomes. A plethora of reports can overwhelm leadership, overwhelm the SOFC regarding report generating, and be considered ineffective if they are never used or reviewed. SOFC reports should be used to help create effective security improvements, not just produced for the sake of producing reports. Cybersecurity leaders must realize the importance of taking a systematic and data-driven approach to enabling continuous improvement for the SOFC. The first step in this metric reporting process is to define clear and measurable goals and objectives. This provides a roadmap and a basis for evaluating performance over time. Regular assessments of security operations processes are critical for identifying areas for improvement and staying ahead of changing threats and technology. Utilizing threat intelligence and analytic metrics can help the SOFC and cybersecurity leadership stay ahead of the latest threats and proactively defend against attacks. Threat intelligence and analytics helps the SOFC to quickly identify, analyze, and respond to threats in real-time using advanced tools and technologies.

In addition to these technical reporting strategies, it is equally as important to foster a culture of collaboration and continuous learning within the SOFC. This can be achieved by encouraging open communication, collaboration, and knowledge sharing among team members, as well as providing opportunities for professional development and training. Regular customer feedback is essential for identifying areas of the SOC's operations that need improvement. This feedback should be analyzed, prioritized, and addressed in a timely manner to ensure the SOC is providing the best possible service. Finally, embracing new technologies and tools is crucial in the rapidly changing cybersecurity field, as they can improve operations and better protect customers.

In this chapter, we did a deep dive into the critical topic of cybersecurity reporting. We emphasized the importance of leveraging effective reporting as a tool to drive security improvements within an organization. The chapter stressed the need for tailored and leveraged reporting, with a focus on appropriate reporting frequency. This chapter further highlighted the idea that few cybersecurity leaders fully grasp the potential impact that effective reporting can have on the security posture of their organization. To maximize the benefits of reporting, it is crucial to avoid generating excessive reports for the sake of appearance. Instead, the emphasis should be on creating reports that drive real security improvements and support the achievement of desired outcomes. Effective reporting should be viewed as a key aspect of promoting security, not a burdensome task to be disregarded.

9
SOFC Metrics

Metrics are the voice that speak the tale of your cybersecurity story, be it good or bad; they provide a comprehensive picture of your work.

Dr. Kevin Lynn McLaughlin, PhD

Metrics Tell the Story

As you can imagine, maintaining a security posture that has more than 100 active botnets at any given time is not a recipe for success for any chief information security officer or cybersecurity team. However, at one company I worked with, this was the norm, and it went completely unnoticed. To make matters worse, when the team was asked if they had any cybersecurity concerns or issues, the standard answer was "no, we do not." That answer wasn't a lie because they truly believed that. They believed it because they had no reporting or metric program in place to monitor, detect, respond, and remediate issues of concern. In the world of cybersecurity, ignorance is not bliss; it is simply a way of mentally avoiding the serious attacks and issues that are taking place. There are two powerful tools in any chief information security officer's arsenal. One is the ability to put a Cybersecurity Operations and Fusion Center (SOFC) in place to increase visibility of the security posture of an organization. And the other powerful tool is being able to show quick dramatic improvements in handling the cybersecurity basics. This chapter will highlight some of the ways that the reader can use to create powerful metrics that clearly show the cybersecurity benefits of having a highly effective SOFC in place. Executives need to know if the dollars they are providing towards security improvements are being used effectively to build good levels of security improvements. This chapter explains how to tell that story through the power of metrics. Regardless of what you hear others say about metrics being misleading

DOI: 10.1201/9781003259152-12

and subject to manipulation (they can be) that doesn't mean that you cannot make use of legitimate metrics to tell your story. Metrics do matter.

Type and Timing – Metrics That Matter

Metrics play a crucial role in measuring the effectiveness of a SOFC. The right metrics help organizations understand how well they are performing, identify areas for improvement, and make data-driven decisions. The following operational metrics are particularly important for SOFC operations and effectiveness:

- **Time to Detect (TTD)** – This metric measures the time elapsed between when a security incident first occurs and when it is detected by the SOFC. TTD is important because it directly impacts the speed with which threats can be neutralized and the potential damage caused.
- **Time to Respond (TTR)** – TTR measures the time elapsed between when an incident is detected and when it is responded to. TTR is important because it directly impacts the speed with which threats can be neutralized and the potential damage they can cause.
- **Mean Time to Resolution (MTTR)** – MTTR measures the average time it takes to resolve a security incident. This metric is important because it provides an overall measure of the effectiveness of the SOFC and helps organizations identify areas for improvement.
- **False Positive Rate (FPR)** – FPR measures the number of false alarms generated by the SOFC detection systems. This metric is important because it helps organizations understand the accuracy of their detection systems and ensures they are not wasting valuable time and resources responding to false alarms.
- **Threat Detection Rate (TDR)** – TDR measures the number of threats that are successfully detected and neutralized by the SOFC. This metric is important because it helps organizations understand the effectiveness of their threat detection and response processes.

- **Incidents Sent Out to Support Teams** – This includes Windows administrators, desktop support, Linux administrators, dBase administrators, etc. It's also important to include the status of each of the items that are pending resolution. This is an important metric because it helps ensure that items that pose risk to the organization are being remediated as intended.

- **Performance against Service Level Agreements (SLAs)** – This metric measures the quality of an SOFC service delivery in meeting agreed-upon standards of availability, reliability, and responsiveness. It is an important metric because it provides an objective way to assess and improve the SOFC performance and helps ensure that customers receive the level of service they expect.

- **Service Request** – This metric measures the number of requests received by the SOFC for services, such as new cybersecurity automations, hunt hypothesis, Red Team engagement requests, or security configurations. It is important because it provides insight into the SOFC workload, helps prioritize resources, and identifies areas where process improvements may be needed.

- **Change Task** – This metric measures the number of changes made to an SOFC infrastructure or processes, such as software upgrades or policy updates. It is important because it helps assess the impact of changes on the SOFC operations, identifies potential risks, and tracks compliance with change management policies.

- **Problem Tickets** – This metric measures the number of incidents or issues that have been reported to the SOFC. These incidents/issues include security breaches or system failures. These metrics are important because they help assess the SOFC's ability to respond to and resolve issues, identify areas for improvement, and track the effectiveness of incident response procedures.

The following executive level metrics are essential when telling your cybersecurity story to the organization's leaders. The metrics also help establish the foundation of understanding that allows an organization's leaders to support and buy into the cybersecurity program. Effective communication and reporting of SOFC executive level metrics are critical for demonstrating the value and impact of these functions to executive management. The following executive level metrics are impactful for an organization's leadership team:

- **Threat Intelligence Coverage:** Measures the completeness and relevance of the threat intelligence gathered and processed by the SOFC. A high Threat Intelligence Coverage metric indicates a robust and effective threat intelligence function. A low Threat Intelligence Coverage metric may indicate a need for improvement in the threat intelligence collection and analysis processes.

- **Incident Severity:** Measures the impact and criticality of security incidents through the use of a standardized severity rating system. This metric can be used to demonstrate the level of risk and potential harm posed by security incidents, and to prioritize the incident response actions.

- **Threat Actor Trends:** Measures the evolution of the threat actors, their tactics, techniques, and procedures, and the changes in their motivations and goals. This metric can be used to demonstrate the evolving nature of the cyber threat landscape and to inform future incident response planning and strategies.

- **Percentage of Work Being Completed by Automation, AI, and ML:** This metric measures the extent to which cybersecurity operations are relying on automation, artificial intelligence (AI), and machine learning (ML) technologies. It's an important metric to monitor because it provides an objective measure of the adoption and effectiveness of these advanced technologies in enhancing the performance and efficiency of cybersecurity operations. Here are a few additional metrics associated with automation, AI, and ML that could be impactful for organizational executives:
    - Automation Adoption: The level of adoption and integration of automation, AI, and ML technologies into cybersecurity operations, and the extent to which they are being leveraged to improve performance and efficiency.

- Workload Reduction: The impact of automation, AI, and ML technologies on the workload of cybersecurity professionals, and the extent to which these technologies are reducing the manual labor and repetitive tasks involved in cybersecurity operations.
- Accuracy and Consistency: The quality and accuracy of the results generated by automation, AI, and ML technologies, and the extent to which these technologies are improving the consistency and reliability of cybersecurity operations.
- Resource Optimization: The impact of automation, AI, and ML technologies on the allocation and utilization of cybersecurity resources, and the extent to which these technologies are enabling organizations to optimize their cybersecurity investments and capabilities.
- **Time to Detect:** Measures the elapsed time between the occurrence of a security incident and its detection by the SOC. A low TTD metric indicates an efficient incident detection process. A high TTD metric may indicate a need for improvement in the organization's monitoring and detection systems.
- **Time to Respond:** Measures the elapsed time between the detection of a security incident and the initiation of the response process. A low TTR metric indicates a rapid and effective incident response process. A high TTR metric may indicate a need for improvement in the incident response policies and procedures.

Timing of Metric reporting

An SOFC should provide operational metrics on a regular basis, such as daily, weekly, or monthly. The frequency of providing metrics depends on the organization's needs and the pace at which their cybersecurity operations are evolving. Providing metrics on a regular basis is important because it allows the cybersecurity leaders to stay informed and make timely decisions. Real-time visibility into the performance of the SOFC programs is crucial to help ensure potential security threats are detected and responded to in a timely manner. Regular metric reports also provide an opportunity for management to identify trends and patterns and make changes to the programs if necessary. By providing metrics on a regular basis, the organization can ensure its security posture remains up-to-date and their resources are being used efficiently. The timely distribution of metrics also helps foster transparency and accountability and provides a clear picture of the organization's overall security posture. I can't stress it enough that the best timing for providing SOFC operational metrics is on a regular basis, whether daily, weekly, or monthly. This helps ensure that the organization can make informed decisions, stay up to date on the performance of their security operations, and minimize risks. Sending metrics on a regular basis is important because they provide a quantitative and objective measure of the performance and impact of the SOFC functions. These metrics also enable executive management to make timely informed decisions and allocate resources based on the value and performance of these functions. The effective communication and reporting of SOFC metrics are crucial for demonstrating the value and impact of these functions to executive management. It also helps guide future investment and improvement decisions. The best time for providing SOFC executive metrics would be monthly. This schedule provides regular, up-to-date information for executives to make informed decisions while also allowing for trending analysis over time. Additionally, monthly metrics allow for timely identification and resolution of potential security issues. Lastly, metrics should be provided to the Board of Directors on a quarterly basis. This schedule strikes a balance between providing frequent updates and not overwhelming the board with too much information. A quarterly cadence allows for a comprehensive overview of the organization's security posture and provides sufficient time for the board to review and act on the information. Furthermore, it aligns with the typical reporting cycle of most organizations, ensuring that security metrics are integrated into the broader reporting landscape.

It's important to incorporate both operational and executive level metrics when telling the story about the performance of the SOFC program for a global organization. Operational metrics, such as the percentage of work being completed by Automation, AI, and ML, are important because they provide a granular view of the day-to-day performance of the SOFC programs. These metrics help organizations understand the impact of their cybersecurity operations on the workload and efficiency of their security teams, as well as the quality and accuracy of the results generated by automation and other advanced technologies. Executive level metrics, on the other hand, provide a high-level view of the performance of the SOFC programs, and are essential for communicating the value and impact of these programs to executive management and other stakeholders. Examples of executive level metrics include the total cost of cybersecurity, the TTD and respond to cyber threats, and the number of data breaches and incidents. By regularly tracking and analyzing both operational and executive level metrics, organizations can gain a comprehensive view of the performance of their SOFC programs and make informed decisions about their security operations. This enables organizations to identify areas for improvement, prioritize their cybersecurity investments, and ensure they are well-prepared to detect and respond to threats in real-time. It also helps organizations minimize risks and maintain the highest standards of cybersecurity. The combination of operational and executive level metrics provides a comprehensive and compelling story about the performance of the SOFC programs for a global organization and enables organizations to make informed decisions about their security operations and investments.

In summary, this chapter focused on the importance of metrics in cybersecurity. Metrics are an effective tool for telling the story of the organization's security posture and the impact of security improvements. The chapter emphasizes the importance of both operational and executive metrics, each of which serves a different purpose. Operational metrics provide insight into the day-to-day operations of the security program, while executive metrics give a high-level overview of the organization's security posture for decision-makers. The proper timing for sending metrics is essential for making necessary program adjustments in a timely fashion. This allows for informed decisions to be made and for potential security issues to be promptly addressed. The chapter concludes by highlighting the value of a well-executed SOFC and how metrics can demonstrate the benefits of a highly effective SOFC. Metrics are an effective and important tool in any cybersecurity/SOFC continuous improvement program.

PART IV
LEADERSHIP ALIGNMENT AND SUPPORT

10
SOFC ALIGNMENT
AND SUPPORT

Much like the Titanic's noble ambition to journey from point A to point B, a cybersecurity program will suffer a similar fate if not fortified with resolute backing and dedication from the organization's executive leadership team.

Dr. Kevin Lynn McLaughlin, PhD

Gaining Leadership Alignment and Support

Successful organizational endeavors need to have alignment and buy-in from the organization's executive leadership team. Building and spending money on a Cybersecurity Operations & Fusion Center (SOFC) is not an exception to this rule. This chapter discusses tips, techniques, and strategies for obtaining executive buy-in and support for the SOFC. We will dive into clear examples of how an effective SOFC can drive security improvements that matter. These examples can be shown to executives to gain support and to help show them why monitoring, detection, and response capabilities are critical to an organization's fight against threat actors.

Cost Benefit

An organization's SOFC can be cost-effective because it allows for efficient and centralized management of cybersecurity threats and incidents. An SOFC provides real-time monitoring and response to security incidents. It also provides valuable insights into potential security threats and helps organizations better allocate their security resources. Organizations can more quickly detect, respond to, and mitigate security threats. This reduces the impact of these threats

DOI: 10.1201/9781003259152-14   **87**

and the overall cost of cybersecurity by having an SOC and Fusion Center in one place. Having a dedicated SOFC can reduce the need for multiple security tools, process, and methodologies resulting in cost savings for the organization. An SOFC provides real-time monitoring and response to security incidents. This allows organizations to detect and respond to threats before those threats cause severe damage. This reduces the cost of cleanup and recovery after a security breach. Organizations can more efficiently allocate their security resources and reduce the cost of various security tools by centralizing their cybersecurity operations within the SOFC. The SOFC can also provide valuable threat intelligence to help organizations better understand the types of threats they are facing. This allows an organization to make more informed security investments. SOFCs are also great for increasing overall cybersecurity efficiency. The SOFC can automate many routine security tasks, which frees up security personnel to focus on higher-level tasks. This can ultimately reduce the cost of hiring and training additional staff. Another technique that improves the cost effectiveness of your SOFC is making use of a risk-based approach to prioritize and allocate resources. This approach involves assessing the potential impact and likelihood of cyber threats and allocating resources to mitigate the highest-priority risks. Leveraging cybersecurity and cyber threat automation, along with artificial intelligence and machine learning tools, can reduce manual effort, increase efficiency, and reduce costs. Using a solution that integrates with existing security tools and infrastructure can help reduce the need to duplicate work streams and improve overall security posture – also resulting in cost savings. When all is said and done, a cost-effective SOFC should be risk-focused, automated, and integrated.

An in-house SOFC can be cost-effective by reducing reliance on third-party vendors, leveraging existing technology and staff, and improving efficiency through specialized training and streamlined processes. Having a dedicated team in place can improve response times to security challenges that arise. That team can also provide knowledge of the company's specific security needs – which can lead to more targeted solutions. Additionally, proper planning and resource allocation can help minimize costs and maximize return on investment. You need to spend time analyzing the following factors when thinking about creating an in-house SOFC:

- **Personnel Costs:** Calculating the cost of hiring and training security professionals is critical to determining the viability of an in-house SOFC. This includes the salaries and benefits of the security staff and the costs associated with their training and development. In addition to direct costs, organizations should consider the indirect costs associated with hiring, such as a budget for recruiting and onboarding new employees.
- **Equipment and Infrastructure Costs:** Assessing the costs of hardware, software, and other technology required to run the in-house SOFC is essential. Organizations should consider the upfront costs of acquiring these technologies, as well as the costs associated with their maintenance and upgrades over time.
- **Maintenance Costs:** Ongoing maintenance costs for hardware, software, and security infrastructure must also be taken into consideration. This includes costs for regular updates, upgrades, and other ongoing maintenance needs that are essential for keeping the in-house SOFC operational and secure.
- **Response Time:** Lastly, it is important to consider the time required to respond to security incidents. A slower response time can result in more costly breaches. Therefore, organizations must carefully evaluate their ability to respond to cybersecurity incidents in a timely and effective manner. This requires a comprehensive understanding of the skills and resources required to maintain an in-house SOC, as well as a realistic assessment of the organization's ability to respond to security incidents.

These data points must be carefully evaluated and considered when deciding whether to create an in-house cybersecurity SOFC facility. A well-informed decision can help organizations determine the most cost-effective and secure solution for their specific needs.

Selecting a managed services solution for an SOFC can be cost-effective because it allows an organization to off-load the cost and resources associated with building and maintaining an in-house SOFC through a third-party provider. These off-loaded costs include lower overall expenses for technology, staffing, training, and expertise. Additionally, managed services providers often have economies of scale and can provide a higher level of security for a lower cost compared to an in-house solution. Managed service providers also offer access to a wider range of security tools and expertise, reducing the need for in-house investment in technology and staffing. To sum it up, an SOFC run by a managed service solution can provide a cost-effective alternative for organizations seeking to improve their cybersecurity posture. Consider the following factors when completing your analysis:

- **Outsourcing Costs:** This includes the costs of outsourcing security operations to a third-party service provider, such as ongoing service fees, licenses, and support. It is important to determine the cost of these services to compare them to the cost of in-house security operations.
- **Expertise:** Evaluate the level of expertise offered by the managed service provider and compare it to the expertise of your in-house security staff. To ensure your organization's security needs are met effectively, the expertise of the managed service provider should be at par or better than your in-house staff.
- **Scalability:** Assess the ability of the managed service provider to handle growth and changing cybersecurity needs over time. This is important because as your organization grows, your cybersecurity needs will change. The managed service provider should be able to adapt to these changes effectively.
- **Incident Response:** Consider the response time and expertise of the managed service provider in responding to security incidents. The ability of the managed service provider to respond quickly and effectively to security incidents is crucial in minimizing damage to your organization.
- **Availability of Resources**: To ensure adequate support, it's important to evaluate the resources available to the managed service provider, such as technology, infrastructure, and personnel. The availability of resources is essential to ensure that the managed service provider can provide the necessary support and expertise for your organization's security needs.

When deciding whether to use a managed services partner for developing your cybersecurity SOFC, it is important to carefully consider each of these above data points to make an informed decision.

It's important to do your due diligence when deciding whether to use an in-house SOFC versus a managed services SOFC. An in-house SOFC offers organizations complete control over their security operations and the ability to tailor the solution to their specific needs. However, it also requires significant investments in personnel, technology, and infrastructure, as well as ongoing maintenance costs. In addition, organizations must be prepared to dedicate the necessary resources to effectively manage and maintain their in-house SOFC. On the other hand, a managed services SOFC provides organizations with a cost-effective alternative by outsourcing the management and maintenance of their security operations to a third-party provider. This can result in lower overall expenses for technology, staffing, training, and expertise, and can provide access to a wider range of security tools and expertise. Managed services providers often have economies of scale and can provide a higher level of security for a lower cost compared to an in-house solution. At the end of the day, organizations must carefully evaluate their specific needs and resources when deciding between an in-house SOFC and a managed services SOFC. An in-house SOFC may be the best choice for organizations with an elevated level of technical expertise and the resources to invest in personnel, technology, and infrastructure. Conversely, a managed services SOFC may be a more cost-effective solution for organizations seeking to improve their cybersecurity posture without a significant investment in personnel and technology. Regardless of approach, each of these solutions should be put in place with one to two non-shared resources who are dedicated to ensuring that your SOFC is working as intended. You want resources whose attention is focused on your capabilities and how they are working and not those whose attention is spread across multiple customers, of which you may be the smallest fish in that pond.

Executive Buy-in

For any program to be successful, executives within the organization need to buy into the idea of that program and its importance to the bottom line of the business. Here are some ideas for securing executive buy-in for your SOFC.

- **Articulate the Business Value:** It is essential to communicate the tangible business benefits of a robust cybersecurity SOFC program. The benefits can include improved security posture, reduced risk of data breaches, and increased efficiency in incident response. Demonstrating the potential fiscal impact of a breach and the cost savings of proactively addressing security risks can help gain executive support.
- **Show Evidence of Industry Best Practices:** Highlighting industry best practices and demonstrating how similar organizations have successfully implemented similar programs can help build credibility and gain executive buy-in.
- **Build a Strong Business Case:** Developing a comprehensive business case that includes a detailed budget and implementation plan can help demonstrate the feasibility of the program and secure the necessary resources and support from executives.
- **Foster Cross-Departmental Collaboration**: Collaborating with other departments, such as IT, Legal, and HR, can help build a more comprehensive security/cybersecurity program and demonstrate the importance of this across the organization.
- **Involve Executives in the Process:** Inviting executives to participate in key meetings and decision-making processes can help build buy-in and ensure their active engagement and support for the program.
- **Continuously Communicate Results and Progress:** Regularly communicating the results and progress of the SOC and threat intelligence center programs to executives can help demonstrate the ongoing value and impact of the initiatives.

Sharing cybersecurity war stories is another way of securing executive buy-in and funding. It provides the executives with real-life examples of how the team saved the company from serious cyber incidents. These stories work well to show a need for the SOFC by highlighting real-world threats and incidents that the SOFC has dealt with. The SOFC can demonstrate the tangible benefits of its work and the value of the organization's investment by sharing specific examples of deterred threats and averted incidents. Telling these stories in a way that does not introduce fear, uncertainty, and doubt helps build executive trust and helps executives see the tangible results of the SOFC's work. Executives are more likely to trust and support the SOFC when they can see that their investment is being put to effective use. The thoughtful use of cybersecurity war stories helps raise awareness about the importance of cybersecurity and encourages other departments and stakeholders to prioritize cybersecurity in their own decision-making processes. Consider having a standard script established for executive tours of the SOFC. Offering a tour of the cybersecurity SOFC can help generate executive buy-in for the program. This allows executives to see the processes and tasks involved with the SOFC first-hand. This hands-on approach can give executives a better understanding of the importance of the program and its potential impact on the organization. Additionally, demonstrations of the program can highlight its capabilities and effectiveness. This helps build confidence and support for the initiative among key decision-makers. By providing a clear and tangible representation of the program, executives are more likely to understand the need to invest in and support the initiative. When touring executives through the SOFC, it's important to highlight the potential that AI technologies have to enhance the capabilities of the SOFC. Be creative in developing demonstrations that highlight the use of AI technologies, such as building SOAR scripts or threat hunting bots to actively search for potential threats in the organization's infrastructure around-the-clock. By highlighting the advanced capabilities of the program, executives can see the true potential of the cybersecurity SOFC and the benefits it can bring to the organization. Demonstrating how these technologies are constantly working to detect and respond to potential threats can help build support and buy-in from key decision-makers. By displaying the innovative capabilities of the program, the tour can help establish the cybersecurity SOFC as a critical component of the organization's overall security strategy.

It is essential to align the SOFC program with the organization's business objectives. It's also important to provide a risk-based approach that demonstrates a detailed understanding of the current threat landscape and how the program will address specific risks. Proactively quantifying the cost of a breach can help increase an executives' willingness to invest in security initiatives. Engaging with executives regularly through check-ins and updates can help build and maintain their ongoing support for the SOFC. Lastly, having a dedicated and skilled team in place to implement and manage the SOFC program can increase executive confidence in its ability to deliver results. By following these best practices, organizations can work towards securing the necessary resources and support to successfully implement and sustain their cybersecurity SOFC and other cybersecurity programs.

Metrics That Tell the Story for You

Using interesting and exciting stories that are based on a foundationally well-planned set of metrics can help highlight the impact the SOFC has had on an organization. This can be done by emphasizing that the SOFC brings increased efficiency and effectiveness to the organization's overall cybersecurity program. Emphasizing the SOFC's ability to respond to and mitigate potential threats in a timely manner can also demonstrate its value to the organization. Telling success stories of the SOFC helps paint a picture of the program's overall impact on the organization. These stories should be communicated to executives and stakeholders – which can help secure ongoing support for the initiative.

In conclusion, this chapter explored the topic of gaining leadership alignment and support for the SOFC program. We highlighted the significance of gaining the support of the organization's executive leadership team. This is crucial for the success of the SOFC program. The chapter covered various tips, techniques, and strategies to obtain executive buy-in and support. This includes displaying the cost benefits of the program, the importance of having a fast time to detect, the value of having expert response capabilities, and the power of metrics in telling the story of the program's success. The SOFC program is crucial in the fight against cyber threats and attacks. It is vital for organizations to have the resources and support to implement an effective SOFC program. This chapter also offered a comprehensive overview of the steps that can be taken to gain leadership alignment and support. By demonstrating the value of the program to the executive leadership team, organizations can secure the necessary support and resources to effectively mitigate cyber threats and improve their overall security posture.

11

KEY COMPONENTS OF A TURNKEY SOFC

Strategic partnerships in cybersecurity are not a one-size-fits-all solution. Timing and circumstances play a crucial role in determining the appropriateness of seeking outside support for managing a cybersecurity program.

Dr. Kevin Lynn McLaughlin, PhD

What to Look for When Buying a Turnkey SOFC Solution

Not all organizations have the desire, resources, or the capability to build their own Cybersecurity Operations and Fusion Center (SOFC). In these cases, it may be beneficial for organizations to seek out a turnkey solution. This type of solution provides a comprehensive package of services and support that can effectively address the organization's security needs. When evaluating a turnkey SOFC solution, it is essential to consider a variety of factors to ensure it is a good fit for your organization. These factors include planning the work to be done, ensuring that the solution integrates well with your organization's processes, establishing effective escalation procedures, and having access to daily, weekly, and monthly metric reporting. These metrics can help you track the effectiveness of the solution and make informed decisions about your organization's security posture. Considering these factors can help organizations ensure that their turnkey SOFC solution is aligned with their security objectives and will help improve their overall cybersecurity posture. With the right turnkey solution in place, organizations can focus on their core business operations and feel confident that their security needs are being managed.

DOI: 10.1201/9781003259152-15

Organizations that decide to hire a turnkey solution to operate their SOFC should look for a partner that will be invested in the protection of their data and assets and have a passion for cybersecurity. The right managed services partner can provide organizations with the expertise and resources needed to effectively manage their security operations and proactively identify and respond to threats. Organizations should look for several key characteristics when choosing a managed services partner. These characteristics can help indicate a commitment to delivering high-quality services and best practices. First, organizations should seek a partner that has a deep understanding of the current cybersecurity landscape and is well-versed in the latest security technologies and techniques. This partner should be able to provide actionable insights and guidance to help organizations stay ahead of emerging threats. Another important consideration is the partner's history of delivering high-quality services and providing organizations with the resources they need to effectively manage their security operations. Organizations should look for a partner that has a proven track record of delivering effective SOFC services and has the experience and expertise to handle complex security challenges. Organizations should look for a managed services partner that is committed to developing a strong relationship with their clients. This partner should be able to provide ongoing support and guidance to help organizations stay ahead of emerging threats and should be able to work closely with the organization's internal teams to ensure their security needs are met. Organizations can ensure their security operations are well-positioned to proactively identify and respond to emerging threats by choosing a managed services partner that is committed to delivering high-quality services and best practices.

A great managed services team for your SOFC should bring a comprehensive set of tools, processes, and attitudes to the organization. The team should have a well-established set of tools that are used to manage, monitor, and respond to security incidents. These tools should be able to collect and analyze substantial amounts of data from various sources like network logs, endpoint security systems, and security information and event management systems. They should also have the capability to automate certain tasks, such as triage and escalation. This helps ensure that security incidents are addressed in a timely and effective manner. The processes used by the managed services team should be well-defined and well-documented, with clear procedures for incident response, escalation, and communication. The team should be proactive in their approach to security, continuously monitoring for threats and vulnerabilities, and taking action to mitigate them before they cause harm to the organization. The team should have a positive attitude and be passionate about security. They should be dedicated to helping the organization achieve its security goals and be willing to work closely with the organization's internal security teams to ensure they are aligned and working towards the same objectives. When choosing a managed services partner for your cybersecurity SOFC, it is important to look for a team that brings a comprehensive set of tools, processes, and attitudes to the table. A great managed services partner will help ensure that your organization is able to effectively manage its security risks and achieve its security goals. The word partner is important. As a cybersecurity expert with extensive experience in the field, I can say that there is a stark difference between a quality, managed service partner for a cybersecurity SOFC, and a company that merely wants to take your money. A managed service partner that is committed to helping organizations improve their cybersecurity posture will bring a range of tools, processes, and the right attitude to the table. This includes a thorough understanding of the latest cybersecurity threats, as well as the latest best practices for protecting against them. The right partner will also bring a commitment to continuous improvement, and a willingness to work collaboratively with your organization to ensure that your SOFC effectively protects your critical assets. On the other hand, a company that simply wants to take your money will be focused solely on securing a payment and delivering bare minimum service. They will be less concerned with the quality of their work, and may not have the latest tools, processes, or expertise necessary to effectively protect your organization. Additionally, they may be less willing to work with your organization to continuously improve your security posture and may not provide the level of support and guidance that a quality managed service partner would. In short, if your organization is looking for a strong and effective cybersecurity

SOFC, it is essential to work with a trusted managed service partner that is committed to helping you achieve your goals and improve your security posture.

Plan the Work

Even turnkey solutions will require work from your in-house staff to help the SOFC provider understand how to work effectively and efficiently with your organization. With that in mind, it is important that you work closely with your managed services partner. You must have a clear understanding of the program and project planning that needs to take place to build a successful cybersecurity SOFC with a managed service partner. This requires careful consideration of several key factors. First, the organization must have a clear understanding of its security needs, goals, and objectives, as well as the available resources and budget the organization has at its disposal. This will help the managed service partner develop a customized solution that meets the specific requirements of the organization. Next, it is important to establish clear lines of communication between the organization and the managed service partner. This includes the definition of roles and responsibilities, as well as the development of a detailed project plan that outlines the steps required to deploy the SOFC. The full-time employees (FTEs) within the organization will play a significant role in ensuring the success of the managed service partnership. They should be engaged in regular communication with the managed service partner and be prepared to provide support and feedback throughout the deployment process. In addition, the FTEs within the organization should be prepared to work closely with the managed service partner to ensure that the necessary policies, processes, and procedures are in place to support the SOFC operation. This includes providing access to relevant systems and data, as well as the development of detailed metrics and reporting to track the performance of the SOFC. The success of a managed service partnership for a cybersecurity SOFC will depend on a combination of effective communication, collaboration, and the development of a well-structured program and project plan. Taking these considerations into account will help an organization ensure that their managed service partner is successful in deploying a strong and effective security solution that meets the specific needs of their organization.

Integration into Your Organization Processes

I must stress just how important it is to have a comprehensive integration of your chosen managed service SOFC provider into your organizational processes. This integration helps ensure the effective functioning and security of the organization. A fully integrated provider works closely with the organization's processes. This ensures a smooth flow of information and decision-making. It also eliminates operational gaps and duplications of efforts, which leads to more efficient processes. A fully integrated managed services SOFC provider works closely with various departments, including IT, Finance, Legal, and human resources, to ensure a comprehensive and effective security posture. Here are a few ways the SOFC will collaborate with an organization's internal teams:

- **IT Processes:** An SOFC provider works closely with the IT department to ensure the proper integration of the provider's security solutions into the organization's IT infrastructure. This helps the organization identify and mitigate security threats in real-time and assists in ensuring the organization's IT systems remain compliant with relevant regulations and standards. This also allows the managed SOFC to leverage existing IT tools and solutions in order to escalate cybersecurity incidents to the proper resolution owners.
- **Finance Processes:** The SOFC provider works closely with the Finance department to ensure that the organization's budget is distributed effectively for the implementation of security solutions and for the mitigation of security threats. This helps ensure that the organization has the resources necessary to effectively manage its security posture.
- **Legal Processes:** The SOFC provider works closely with the Legal department to ensure that the organization's security policies and procedures follow relevant laws and regulations. This helps reduce the risk of legal penalties and reputational damage.

- **Human Resources Processes:** The SOFC provider works closely with the human resources department to ensure that the organization's employees receive the necessary security training and awareness programs. This helps reduce the risk of security incidents that result from employee actions. The partnership with HR also allows for clear and concise cybersecurity operating procedures to be established when working on sensitive and confidential cybersecurity cases or when assisting HR in their investigations.

Your SOFC provider will work closely with various departments within your organization when it is fully integrated into your organization's cross-functional processes. This will help your organization establish a comprehensive and effective security posture. This helps ensure that the organization's security goals are aligned with its overall business goals. It will also help ensure that your organization's security posture is constantly evolving to meet the changing threat landscape. One of the best practices for integration is to establish clear communication channels between the managed services partner and the organization's cross-functional teams. This helps the managed services provider stay aware of the organization's security needs and requirements and ensures that the organization is aware of the provider's capabilities and limitations. Another important practice is to regularly review and update the security policies and procedures. This is essential because of the constantly evolving threat landscape. It is also crucial to provide the necessary training and awareness programs to the organization's employees to reduce the risk of security incidents resulting from employee actions. This helps create a culture of security within the organization and helps reduce the risk of security breaches. It's important to design clear communication channels, regular policy, and procedure reviews, as well as employee training and awareness programs to ensure a comprehensive integration of a cybersecurity managed services partner. This can help create a stronger security posture, reduce the risk of security incidents, and align the organization's cybersecurity goals with its overall business goals.

The managed service provider's expertise and technology enhance an organization's risk management capabilities. This helps identify and mitigate potential threats in real-time – which ultimately leads to a stronger overall security posture for the organization. It's important that organizations stay compliant to avoid penalties and reputational damage due to the constantly evolving cybersecurity regulations and standards. An integrated provider can assist in ensuring the organization meets the relevant standards and regulations. The provider's access to real-time security data and threat intelligence provides valuable insights to the organization, leading to better informed and effective decision making. The integration of a managed service SOFC provider into an organization's processes is of paramount importance. This integration leads to improved processes, enhanced risk management, better compliance, and more informed decision making, which results in a stronger and more secure organization.

Effective Escalation

An effective operational level escalation process between a cybersecurity managed services SOFC provider and the cross-functional teams it works with is extremely important for a successful managed services SOFC provider. This helps ensure the organization's security posture is strengthened, and that potential security incidents are addressed in a timely and efficient manner. Effective operational level escalation involves establishing clear and well-defined communication channels between the managed services provider and the cross-functional teams. This keeps the provider aware of the organization's security needs and requirements, and helps the organization stay aware of the provider's capabilities and limitations. In the event of a potential cybersecurity incident, the managed services provider should follow established escalation procedures to notify the cross-functional teams and coordinate with them to address the incident. This involves promptly providing relevant information about the incident and its potential impact. It also entails coordinating with the cross-functional teams to develop a plan of action to contain, mitigate, and resolve the incident. In addition to prompt escalation and coordination, effective operational level escalation involves regular post-incident reviews to identify any areas for improvement. Post-incident reviews also allow the team to make any necessary changes to the escalation procedures and processes. This helps ensure that the organization's security posture is continuously strengthened and that the risk of future incidents is reduced.

Executive level escalation is important because it establishes a two-way communication channel for escalations between the managed service provider and the SOFC executive leaders. This helps ensure the organization's security posture is strengthened and that potential security incidents are addressed in a timely and efficient manner at the highest level of the organization. Effective executive level escalation involves establishing clear and well-defined communication channels between the managed services provider and the executive leadership team of the organization. This gives the executive leadership team a much-needed awareness of the organization's security needs and requirements. It also ensures that the provider is aware of the executive leadership team's expectations and priorities. In the event of a critical security incident, the managed services provider should follow established escalation procedures to notify the executive leadership team and coordinate with them to address the incident. This involves promptly providing relevant information about the incident and its potential impact, as well as coordinating with the executive leadership team to develop a plan of action to contain, mitigate, and resolve the incident. The executive leadership team's involvement in the escalation process is crucial for ensuring that decisions related to the incidents are made at the highest level of the organization. Executive involvement helps leadership determine whether the organization's resources are effectively utilized to address the incident. This collaboration also helps make sure that the incident is resolved in a manner that aligns with the organization's overall goals and objectives. Effective executive level escalation between a cybersecurity managed services SOFC provider and the cross-functional teams it needs to work with is crucial for ensuring that potential security incidents are addressed at the highest level of the organization and that the organization's security posture is strengthened. It's important that the escalation process includes clear communication channels, prompt escalation and coordination, and the involvement of the executive leadership team.

Based on my experience, it is imperative for organizations to have regular status update meetings between their cybersecurity leaders and their managed service provider. These meetings serve as an important platform for both parties to keep each other informed of the progress and status of the organization's cybersecurity posture. It also provides an opportunity to discuss and resolve any issues or concerns that may arise. Having regular status update meetings helps foster a strong and collaborative relationship between the organization's cybersecurity leaders and the managed service provider. During these meetings, both parties can exchange ideas, provide feedback, and share best practices that can help improve the organization's overall security posture. Moreover, regular status update meetings provide a comprehensive view of the organization's security posture, including the threats it faces, the risks it is exposed to, and the measures in place to address these risks. This helps ensure that the organization's cybersecurity leaders are aware of the current security landscape, the security posture of the managed service provider, and the steps that need to be taken to further strengthen the organization's security posture.

It is recommended to have status update meetings at least once a quarter between the organization's cybersecurity leaders and the managed service provider. However, the frequency of these meetings may vary depending on the size and complexity of the organization, the level of risks it faces, and its overall security posture. For organizations with high cybersecurity requirements, it may be necessary to have meetings on a bimonthly or monthly basis.

A managed services SOFC solution requires good and effective metrics to keep items on track and running as intended. Metrics provide a quantifiable measure of the performance and effectiveness of the SOFC and helps organizations evaluate the value they are receiving from their managed service provider. A managed service provider should include the following service level agreement (SLA) items in their SOFC services: response time, availability, and resolution time. Common SLAs used between a managed service provider and the organization include:

- **Response Time** – This SLA outlines the time frame in which the managed service provider will respond to a security incident. The response time should consider the severity of the incident, as well as the priority it has been assigned by the organization.
- **Availability** – This SLA outlines the hours of operation of the SOFC, as well as the level of availability of the managed service provider's support team. The availability SLA should

ensure that the managed service provider is able to provide support 24/7, if required.

- **Resolution Time** – This SLA outlines the time frame in which the managed service provider will resolve a security incident. The resolution time should consider the severity of the incident, as well as the priority it has been assigned by the organization.
- **Quality of Service** – This SLA outlines the level of service expected from the managed service provider in terms of the accuracy and completeness of the information provided, the response time to inquiries, and the level of expertise of the support team.
- **Reporting** – This SLA outlines the frequency and format of the managed service provider reports provided to the organization. The reporting SLA should include information on the number of incidents detected and resolved, the average time to resolution, and any other relevant metrics.

These SLA items help organizations understand the level of support they can expect from their managed service provider, as well as the time within which they can expect resolution of any security incidents. Key Performance Indicators (KPIs) are another vital component of a managed service provider's SOFC services. Common KPIs include:

- **Mean Time to Detect:** This metric measures the average time taken to detect a security breach – from the moment it begins, to the moment it is detected.
- **Mean Time to Respond:** This metric measures the average time taken to respond to a security breach – from the moment it is detected, to the moment it is resolved.
- **Threat Detection Rate:** This metric measures the rate at which the SOFC systems detect security threats.
- **False Positive Rate:** This metric measures the rate at which the SOFC systems generate false alerts, which can lead to unnecessary resource allocation and workload.
- **Threat Resolution Time:** This metric measures the time taken to resolve a security threat – from the moment it is detected, to the moment it is fully resolved.

- **Dwell Time:** Dwell Time refers to the time that a threat is present in an organization's network before it is detected and addressed. It is a critical metric for measuring the effectiveness of a cybersecurity SOFC. It is critical because it highlights the time that a threat could potentially be causing damage to an organization's assets and data. A lower dwell time is desirable because it means that the SOFC is detecting and resolving threats more quickly, reducing the potential damage caused by the threat. On the other hand, a higher dwell time indicates that the SOFC is taking longer to detect and respond to threats, leaving the organization vulnerable for a longer period.

These KPIs provide an opportunity to measure the success of the SOFC when it comes to achieving specific objectives, such as detecting and responding to security incidents, reducing the time to resolution, and improving overall security posture.

Operational metrics like the ones we discussed in earlier chapters, *the number of incidents detected and resolved, the average time to resolution, and the number of false positives,* are all critical for evaluating the performance of a managed service provider's SOFC services. These metrics provide organizations with a clear understanding of the impact the SOFC is having on their security posture, as well as the level of service they are receiving from their managed service provider. Metrics play a critical role in evaluating the performance and effectiveness of a managed service provider.

Standard SLA items, KPIs, and operational metrics provide organizations with a quantifiable measure of the value they are receiving from their managed service provider. This data help organizations evaluate the impact of the SOFC on their security posture. These metrics are critical for organizations to understand the level of service they are receiving and to help organizations confirm that they are receiving the best value for their investment in cybersecurity.

In conclusion, seeking a turnkey solution for your organization's cybersecurity SOFC operations is a viable option if you have the budget, but lack the time or desire to manage the operations yourself. However, it is essential to thoroughly evaluate potential providers to ensure they meet your organization's needs. Some key factors to consider include the provider's plan for executing the work, their ability to integrate with your organization's processes, their capacity for effective escalation, and their record for reporting daily, weekly, and monthly metrics. SLAs, KPIs, and other operational metrics are also critical components of a successful partnership with an SOFC provider. Prioritizing these key considerations and ensuring a thorough evaluation of potential partners can help organizations take advantage of the benefits of a turnkey SOFC solution. At the same time, these considerations help minimize the risks of a misaligned or ineffective solution. The biggest item I look for is a positive answer to this question: Are they as passionate as I am about securing my organization's data and assets?

12

CONCLUSION

The pursuit of a secure and protected digital world has only just started. Let us approach this journey with excitement and determination.

Dr. Kevin Lynn McLaughlin, PhD

In my over three decades of cybersecurity experience, I have seen many organizations start their journey towards establishing a Cybersecurity Operations and Fusion Center (SOFC). My experience includes building multiple highly effective cybersecurity programs from the ground up. The journey from having no SOFC in place to a fully operational one is not easy, especially for large Fortune 500 organizations. The process typically begins with a recognition of the need for better cybersecurity operations and threat intelligence capabilities. This may come because of a major security incident, regulatory requirements, or simply recognizing that the current approach is not keeping up with the evolving threat landscape. Once the need is recognized, the organization will often conduct a security assessment to determine the current state of their cybersecurity operations and identify areas for improvement. From this assessment, a plan is developed to build or outsource an SOFC, or add the fusion center component to their existing security operations center (SOC). Next, the organization will begin the process of setting up the SOFC. This can involve hiring or outsourcing a team of cybersecurity professionals, procuring the necessary tools and technologies, and establishing standard operating procedures and workflows. It is important to regularly review and update the SOFC processes, technologies, and personnel to ensure the service remains effective once it's up and running. The regular review and update of KPIs and metrics are critical in ensuring the SOFC provides the desired level of security to the organization. The

DOI: 10.1201/9781003259152-16

journey from no SOFC to a fully operational one can be long and challenging. However, SOFCs are essential for helping any organization keep up with the evolving threat landscape and maintain the security of their assets. With careful planning, implementation, and regular review and updates, a Fortune 500 organization can successfully establish an SOFC that provides the necessary level of security for their organization.

I have had the opportunity to observe numerous organizations without a SOC and/or fusion center. The results can be staggering. Without a solid plan in place, these companies can be and often are blind to the presence of hundreds of active botnets operating on their servers, millions of undiscovered cybersecurity vulnerabilities, and even successful hacks that go unnoticed. Whenever I am talking to a peer and they mention that they are the exception and that they have noticed no cybersecurity attacks against their organization, I know that I am talking to someone who just isn't looking. There can be a variety of reasons for this lack of insight but in today's hyper cyberattack environment it simply is not possible, with extremely few examples, that your systems are not under active attack. In today's increasingly complex and dangerous online landscape, it is vital that businesses take proactive measures to safeguard their operations and protect their assets. The first line of defense in this effort begins with the individuals staffing their SOFC. These are the defenders that are establishing the strong foundation upon what your cybersecurity rest. Their efforts can mean the difference between a thriving business and one at risk of becoming yet another statistic.

Building a combined cybersecurity SOC and fusion center, aka an SOFC, can be a game changer for your organization's ability to have an effective cybersecurity program. I know that many chief information security officers and cybersecurity teams may think the cost of such a setup is too high. That simply isn't the case. The truth is that the total cost of ownership for a well-established SOFC is often far lower than you might think. Let's talk about the return on investment for a minute. Sure, there may be an initial investment required to set up your SOFC. But the benefits of having a fully functional and highly skilled team monitoring your network 24/7 cannot be overstated. By having an SOFC in place, you are proactively mitigating or minimizing the risk of a successful cyberattack. Remember, the costs associated with such an attack can be staggering. Think about it: One major cyberattack could result in hundreds of thousands, if not millions of dollars in damages. Not to mention the long-term impact on your organization's reputation. The cost of establishing and maintaining an SOFC pales in comparison. So what is the call-to-action here? If your organization does not already have an SOFC in place, it is time to consider making the investment. The peace of mind that comes with knowing your network is being monitored and protected around the clock is worth its weight in gold. Do not let the fear of costs hold you back from taking this critical step towards securing your organization's future. Any SOFC that you create and deploy will need your oversight and training to ensure that it is working as you intend. This is not a fire and forget scenario, even if you go down the managed services route. If you stand up your SOFC in a low-cost area, and you invest the time and energy to train the team members and ensure they have the tools necessary to do the job you want them to do, you can have an effective and cost-effective team up and running within six months, big company, or less if you have a smaller organizational footprint. The good news is that there are many good managed service partners standing ready to help you achieve your goals and get your organization protected as it should be. The journey and the outcome of the journey is well worth the effort and time it takes to accomplish.

This was a fun journey exploring SOFCs and their significance for organizations of all sizes. Hopefully, the information you read and the time you invested provided you with a comprehensive definition of an SOFC and its value in safeguarding data, assets, and company operations. The pros and cons of purchasing a turnkey SOFC solution versus building an SOFC were discussed, along with a guide to finding the right partner to help establish, deploy, and run the SOFC. This book covers key considerations such as the type of security analyst (dedicated vs. pooled), location of operations (US vs. overseas), and the importance of establishing expectations and background. The reader is advised to select a partner that is flexible and does not force contract changes for every nuance encountered while performing the work. Our journey dove into the process of staffing the SOFC with the appropriate knowledge levels and leaders, including recruitment, team building, and retention strategies. The infrastructure and toolset needed for SOFC success were discussed, such as cloud-first strategy, physical security, and virtual tools and technologies. Emphasis was put on the significance of creating an effective team culture – which consists of maximizing team performance and promoting long-term retention within the company. We also explored the various components and tools required for monitoring and detecting activity across the infrastructure and outlined the crucial roles necessary for SOFC success. We did a deep dive into the available options for setting up and operating the SOFC – which led us to consider factors such as remote versus on-premises workforce, and the impact of COVID-19. We highlighted the importance of developing proper standard operating procedures for running the SOFC. We discussed in detail the day-to-day operations of the SOFC, including the duties and responsibilities of analysts and team leaders. The book explored key roles and duties that set the SOFC apart, such as fusion center operations, threat hunting, Red Team and Purple Team, as well as effective security vulnerability management.

We also emphasized the significance of effective incident response in cybersecurity and provided options for supporting key security tools. Throughout this book we discussed details surrounding effective response for various incident sizes and situations. How to staff and operate a cyber incident response team was another critical topic that we touched on. We discussed the impact that effective reporting can have on enhancing security within an organization and explored specific metrics that are helpful when communicating cybersecurity improvements. The importance of securing executive support was also discussed, along with tips and strategies for achieving this. This book provided best practices when it comes to considering the pros and cons of how to deploy an SOFC before making a final decision. We discussed deploying an SOFC in-house, which involved dedicating internal resources and manpower towards building and maintaining the framework. We also discussed the services of a turnkey SOFC service provider, which entailed outsourcing the deployment and maintenance of the SOFC. In addition to the options presented, the book also shed light on the cost implications of each approach. We even walked through a comprehensive analysis of the expenses involved in deploying an SOFC in-house. This can help organizations make an informed decision based on their budget and manpower constraints. Moreover, the book presented an insightful overview of the costs associated with outsourcing the deployment of an SOFC to a third-party service provider. This allows organizations to compare and contrast the expenses involved in both approaches.

In conclusion, I hope this book took you on a journey that served as a valuable resource for explaining the importance of deploying an SOFC to strengthen your organization's security posture. By providing a well-rounded analysis of the different approaches and the costs involved, this book hopefully enables you to make informed decisions that align with your organization's goals and priorities. I trust that you found this book to be both insightful and enlightening. I look forward to creating more content that educates leaders in the industry on how to better protect their organizations. I'm equally excited about watching you take on future endeavors in the cybersecurity field.

TEMPLATES AND RESOURCES

Figure A.1 is an example of what a daily morning report from the Cybersecurity Operations and Fusion Center (SOFC) could look like.

Figure A.1   Daily SOFC report.

Figures A.2 and A.3 are examples of a threat intelligence report from the SOFC. The last section of this contains remediation recommendations.

Global Security Services
"Results speak louder than words"

TSA Version 1.8.2	**Security Notification Report (SNR)**

Notification Tracking

Notification	GSO_SNR-SB/4888-2020
Reporting	XXXXXX GSO
Report Date	31 January 2023
Reporter	XXXXXXX
Reporter Tel	XXXXXXX
Reporter Email	XXXXXXX
Intel Reviewer 1	XXXXXXX
Intel Reviewer 2	XXXXXXX
Intel Rep/Rev Code	XXXXXXXX
Alternative contact	XXXXXXX

Stats

Threat type	Vulnerability
Vendor Impacted	Other/Multiple (See Description)
Vendor Rating	No Rating/NA
Impact Level	MEDIUM

OJD	NI
SCG	NI
Threat Intel Alliance	

LOW
Risk Level

Title

Flaws Affecting Industrial Control Systems from Major Manufacturers

RISK ANALYSIS

		Weighted Score	Impact Level
Does this VULNERABILITY XXXXX assets ?	NO	1	LOW
Is there an exploit available, Object exploited or under exploit ?	YES	5	HIGH
Is there a Fix, Patch or a temp solution available ?	YES	1	LOW

Description

The U.S. Cybersecurity and Infrastructure Security Agency (CISA) published four Industrial Control Systems (ICS) advisories, calling out several security flaws affecting products from Siemens, GE Digital, and Contec.

CVE IDS		CVSS Score	8.5

CVE-2022-44456	CVE-2023-22331	CVE-2023-22334	CVE-2023-22373
CVE-2023-22339	CVE-2022-45092	CVE-2022-2068	CVE-2023-35256
CVE-2022-2274	CVE-2022-46732	CVE-2022-40267	

Details/ Behaviour Analysis

Summary
The most critical of the issues have been identified in the following products.
Contec:
Vulnerability Overview
CVE-2022-44456: OS COMMAND INJECTION:
CONPROSYS HMI System versions 3.4.4 and prior are vulnerable to an OS Command Injection, which could allow an unauthenticated remote attacker to send specially crafted requests that could execute commands on the server.

CVE-2023-22331: USE OF DEFAULT CREDENTIALS:
In CONPROSYS HMI System Ver.3.4.5 and prior, user credential information could be altered by a remote unauthenticated attacker.

CVE-2023-22334: USE OF PASSWORD HASH INSTEAD OF PASSWORD FOR AUTHENTICATION:
In CONPROSYS HMI System Ver.3.4.5 and prior, user credentials could be obtained via a machine-in-the-middle attack.

CVE-2023-22373: CROSS-SITE SCRIPTING:
In CONPROSYS HMI System Ver.3.4.5 and prior, an arbitrary script could be executed on the web browser of the administrative user logging into the product. This could result in sensitive information being obtained.

CVE-2023-22339: IMPROPER ACCESS CONTROL:
In CONPROSYS HMI System Ver.3.4.5 and prior, a remote unauthenticated attacker could obtain the server certificate, including the private key of the product.

Siemens:
Vulnerability Overview
CVE-2022-45092: PATH TRAVERSAL:
An authenticated remote attacker with access to the affected product's web-based management (443/TCP) could potentially read and write arbitrary files to and from the device's file system. An attacker could leverage this to trigger remote code execution on the affected component.

Figure A.2   SOFC threat intelligence report.

Affected Products and Versions:
SINEC INS: versions prior to V1.0 SP2 Update 1
The following versions of CONPROSYS HMI System (CHS), are affected:
CVE-2022-44456
CONPROSYS HMI System (CHS): Ver.3.4.4 and prior
CVE-2023-22331, CVE-2023-22334, CVE-2023-22373, CVE-2023-22339
CONPROSYS HMI System (CHS): Ver.3.4.5 and prior
GE Digital reports these vulnerabilities affect the following Proficy Historian product:
Proficy Historian v7.0 and higher versions
The following Mitsubishi Electric MELSEC products are affected:
MELSEC iQ-F Series with serial number 17X**** or later:
FX5U-xMy/z x=32,64,80, y=T,R, z=ES,DS,ESS,DSS: Versions 1.280 and prior
FX5UC-xMy/z x=32,64,96 y=T, z=D,DSS: Versions 1.280 and prior
MELSEC iQ-F Series with serial number 179**** and prior:
FX5U-xMy/z x=32,64,80, y=T,R, z=ES,DS,ESS,DSS: Versions 1.074 and prior
FX5UC-xMy/z x=32,64,96 y=T, z=D,DSS: Versions 1.074 and prior
MELSEC iQ-F Series FX5UC-32MT/DS-TS, FX5UC-32MT/DSS-TS, FX5UC-32MR/DS-TS: Versions 1.280 and prior
MELSEC iQ-R Series R00/01/02CPU: All versions
MELSEC iQ-R Series R04/08/16/32/120(EN)CPU: All versions

Note: No known public exploits specifically target these vulnerabilities. These vulnerabilities are exploitable remotely. These vulnerabilities have a low attack complexity.

Remediations and Recommendations:
1. Contec recommends users update to CONPROSYS HMI System (CHS) Ver.3.5.0 or later.
2. CISA recommends users take defensive measures to minimize the risk of exploitation of this vulnerability. Specifically, users should:
3. Minimize network exposure for all control system devices and/or systems, and ensure they are not accessible from the Internet.
4. Locate control system networks and remote devices behind firewalls and isolate them from business networks.
5. When remote access is required, use secure methods, such as Virtual Private Networks (VPNs), recognizing VPNs may have vulnerabilities and should be updated to the most current version available. Also recognize VPN is only as secure as its connected devices.
6. CISA reminds organizations to perform proper impact analysis and risk assessment prior to deploying defensive measures.
7. Siemens released V1.0 SP2 Update 1 for SINEC INS and recommends updating to the latest version.
8. CISA recommends users take defensive measures to minimize the risk of exploitation of these vulnerabilities. Specifically, users should:
9. Ensure the least-privilege user principle is followed.
10. Minimize network exposure for all control system devices and/or systems, and ensure they are not accessible from the Internet.
11. Locate control system networks and remote devices behind firewalls and isolate them from business networks.
12. CISA reminds organizations to perform proper impact analysis and risk assessment prior to deploying defensive measures.
13. CISA also recommends users take the following measures to protect themselves from social engineering attacks:
14. Do not click web links or open attachments in unsolicited email messages.
15. Refer to Recognizing and Avoiding Email Scams for more information on avoiding email scams.
16. Refer to Avoiding Social Engineering and Phishing Attacks for more information on social engineering attacks.

Figure A.3   Second page of the SOFC threat intelligence report.

Figures A.4–A.6 are examples of a standard operating procedure (SOP). The SOFC should be working from core sets of SOPs.

Document number: XXXXX
Name: Alerts Investigation
Revision: 1.0

Work instruction

Table of contents

Figure A.4   SOFC SOP page 1.

Document number: XXXXX
Name: Alerts Investigation
Revision: 1.0

Work instruction

1. Purpose

1.1. The purpose of this document is to know the process of investigating DHS alerts and collect the details of the IP's and initiating further remediation.

2. Scope

2.1. This document applies to all Security Analysts in Security Team.

3. References

3.1. Internal references

N/A

3.2. External references

N/A

4. Definitions

4.1. Locally defined terms

4.1.1. Vulnerability: A vulnerability is a weakness or flaw that can be exploited by cybercriminals to gain unauthorized access to a computer system.

4.1.2. POC: Point of Contact is a person or a department serving as the coordinator or focal point of information concerning an activity.

5. Roles and responsibilities

5.1. Incident Response Team: Responsible for investigating and initiating communication with POC to remediate vulnerabilities.

5.2. Vulnerability Management Team: Responsible for verifying/reporting vulnerability scan.

6. Process Flow

7. Work instructions

7.1. Investigating the IP

When Analyst receives details/information of vulnerabilities, they need to check for the IP's information in Vulnerability Management Tool and gather the details on last scan and vulnerabilities found, followed by Global IP allocation sheet from the network team. Inform the leadership team with below details as follows:

- Vulnerability status

Figure A.5 SOFC SOP page 2.

Document number: XXXXX
Name: Alerts Investigation
Revision: 1.0

Work instruction

- IP details
- Analyst must create ticket and reach out to Network Team to get the details for Point of Contact(POC) or Owner.

7.2. If the IP belongs to Company:

- As soon as the analyst get the details of POC or Owner, an SCTASK should be created of Priority 3 (P3) for POC with the vulnerability details requesting them to remediate and Send an email on the same.
- Analyst must do follow-ups to track the status
- Provide an update to Leadership Team with ticket and initial details.

7.3. If the IP doesn't belong to Company:

Analysts need to check for the IP's information in Qualys and if they find out that IP's doesn't belong to Company,

- Crosscheck with Cyber Risk Analytics & Security Ratings team if the public IP is being reported as Company public IP, so that it can be removed from their records.
- No actions will be taken if Ip doesn't belong to company.
- Security Team need to report the same to Leadership Team.

8. Revision history

Revision number	Description	Reason (CR/CN)
1.0	Initial release	TBD

Figure A.6   SOFC SOP page 3.

Index

Printed in the United States
by Baker & Taylor Publisher Services